War Finance and Logistics in Late Imperial China

Monies, Markets, and Finance in East Asia, 1600–1900

Edited by

Hans Ulrich Vogel

VOLUME 5

The titles published in this series are listed at brill.com/mmf

War Finance and Logistics in Late Imperial China

A Study of the Second Jinchuan Campaign (1771–1776)

By

Ulrich Theobald

BRILL

LEIDEN • BOSTON
2013

Cover illustration: Detail from the copperplate engravings to the second Jinchuan campaign (*Pingding Liang Jinchuan desheng tu*), plate "Conquest of the area around Ripang" (*Gongke Ripang yi dai*). The detail shows Manchu bowmen and musketeers to the right, in close combat with Jinchuan "rebels" hidden in a trench. To the left side and in the back, war towers can be seen. Courtesy of the Staatsbibliothek zu Berlin, Preußischer Kulturbesitz.

Library of Congress Cataloging-in-Publication Data

Theobald, Ulrich.
War finance and logistics in late imperial China : a study of the Second Jinchuan Campaign (1771–1776) / by Ulrich Theobald.
p. cm. — (Monies, markets, and finance in East Asia, 1600–1900)
Includes bibliographical references and index.
ISBN 978-90-04-25310-0 (hardback : alk. paper) — ISBN 978-90-04-25567-8 (e-book) 1. Aba Zangzu Qiangzu Zizhizhou (China)—History, Military—18th century. 2. War finance—History—18th century. 3. War, Cost of—History—18th century. 4. Logistics—History—18th century. I. Title. II. Title: Study of the Second Jinchuan Campaign (1771–1776).

DS797.77.A63T44 2013
951.'5032—dc23

2013019489

This publication has been typeset in the multilingual "Brill" typeface. With over 5,100 characters covering Latin, IPA, Greek, and Cyrillic, this typeface is especially suitable for use in the humanities. For more information, please see www.brill.com/brill-typeface.

ISSN 2210-2876
ISBN 978-90-04-25310-0 (hardback/paperback)
ISBN 978-90-04-25567-8 (e-book)

This book is printed on acid-free paper.

Children of the Camp are we,
Serving each in his degree;
Children of the yoke and goad,
Pack and harness, pad and load.
See our line across the plain,
Like a heel-rope bent again.
Reaching, writhing, rolling far,
Sweeping all away to war!
While the men that walk beside,
Dusty, silent, heavy-eyed,
Cannot tell why we or they
March and suffer day by day.

Rudyard Kipling, *The Jungle Book*,
Her Majesty's Servants,
Parade-Song of the Camp Animals
(New York: The Century, 1897
[London: Macmillan, 1894]), pp. 302–303.

CONTENTS

LIST OF DIAGRAMS, FIGURES, MAPS AND TABLES

Diagrams

Figures

Maps

Tables

ACKNOWLEDGEMENTS

This study is part of a wider research programme focussed on "Monies, Markets and Finance in China and East Asia, 1600–1900: Local, Regional, National and International Dimensions" that was conducted between 2005 and 2011. The funds sponsoring this study have partially been provided by the Deutsche Forschungsgemeinschaft (DFG; German Research Foundation), to whom I wish to express my sincerest gratitude.

I am deeply indebted to my project supervisor, Professor Hans Ulrich Vogel, who not only incited in me the energy to treat a subject that has hitherto received little attention, but also took the trouble to guide me through the intricate language of the Qing bureaucracy. Without his advice a lot of the sophisticated arguments of the political actors would not have been perceived by me. Similarly I have to thank Professor Achim Mittag, with whom I had many a long session brooding over the Qianlong emperor and his role for the fulfilment of the Qing dynasty's historical mission.

I owe special thanks to Professor Dai Yingcong (William Paterson University, NJ), without whose support this study could not have been written. Not only did she provide me with invaluable hints as to sources and particular questions, she also encouraged me to carry out a deeper analysis of the numerous regulations concerning military expenditure.

I am also indebted to Roger Greatrex (Lunds Universitet) for valuable information on the background of the two Jinchuan wars; to Michael Axworthy (University of Exeter) who clarified to me how camel-guns were used; and to George Bryan Souza (University of Texas) who indefatigably helped me with details about artillery of the Qing era, particularly as far as the methods of casting and the materials used are concerned. My thanks also go to Elisabeth Kaske (Carnegie Mellon University, Pittsburgh), who particularly helped me with queries concerning the *juanna* and *juanshu* contributions.

Likewise I must thank Iwo Amelung (Goethe Universität, Frankfurt), who initiated me into the secrets of the First Historical Archives (*Diyi lishi dang'anguan*) in Beijing. Special thanks go to Christine Moll-Murata (Ruhr Universität, Bochum). She not only helped me understand complicated accounting systems, but also tracked down the relevant regulations for military equipment in the ordinances of the Ministry of Works.

Not last, I must express my gratitude to Ms Gu Jia (Wuxi), who helped me find my way through the labyrinth of bureaucratic documents, as well as all the others whom I kept pestering with my questions, and those who criticized and encouraged me: Virginia Aksan (McMaster University, Toronto), Werner Burger (Hong Kong), Chen Hailian (Trier), Mark Elvin (Oxford), Yu and Kai Filipiak (Universität Leipzig), C. Pat Giersch (Wellesley College, MA), Arturo Giraldez (University of the Pacific, Stockton, CA), David Graff (Kansas State University), Thomas Hirzel (Sechselberg), Laura Hostetler (University of Illinois, Chicago), Elise Kimerling Wirtschafter (California State Polytechnic University, Pomona), Kuroda Akinobu (Tōkyō University), Jane Kate Leonard (University of Akron, OH), Peter Lorge (Vanderbilt University, TN), James Miller (Inverness), Shan Kunqin (Hefei), Felix Siegmund (Universität Bochum), Michael Sikora (Westfälische Wilhelms-Universität, Münster), Kenneth Swope (University of Southern Mississippi, MS), Frank Tallett (University of Reading), Peter Way (University of Windsor, Ontario), Wang Qing (München), and Yang Yuda (Fudan University, Shanghai).

A final thank you goes to Karl A. Klewer who agreed to put this somewhat dry subject matter into decent English.

Tübingen
Spring 2013

NOTES ON TRANSLATION AND TRANSLITERATION

Chinese names, for persons and places alike, have been represented in the usual, modern system of *pinyin*. Apart from a few exceptions this also applies to a large degree to Manchurian names and terms when written in Chinese characters. Place names in the area around Jinchuan have been represented using my own system of transcription based on the Chinese pronunciation, as I had nothing but the Chinese variants at my disposal and the better-known Tibetan names represent only the 'high Tibetan' variant. Moreover, since by no means all names of places and persons are known in their Tibetan written form, I have for the sake of consistency used transcriptions based on Chinese throughout, instead of writing some in one form and some in another. For Mongolian, Turkish (Uyghurian) and Persian names there is a wide range of not standardized transcriptions, of which I chose the one that seemed to me the most often used one.

Imperial China as a highly bureaucratized state traditionally had an astonishing variety of public offices, each with its own particular title. To translate these into English there exist a number of aids. Wherever possible I have used Charles O. Hucker's *Dictionary of Official Titles in Imperial China*, the standard and authoritative work of reference although his translations are not always entirely satisfactory. To quote an example, he renders the title *buzhengshi* as 'provincial administration commissioner', whereas a closer equivalent might be 'provincial treasurer'. Also Brunnert and Hagelstrom in their *Present Day Political Organization of China* provided invaluable help with the plethora of military ranks and military terminology in general. Since this book, as opposed to Hucker, only deals with the late 19th and early 20th century it is more exact, especially regarding the titles of the Manchu elite of the Qing dynasty and the ranks within the different military units. All the same even these authors have left gaps, especially when it comes to the lower ranks. These I have taken the liberty of re-translating according to their context.

This book is intended as a study not only for experts in the military, economic or political history of late imperial China, but also for scholars generally interested into the history of warfare, its organisation, and financial and social repercussions. Chinese characters have therefore been banished to the appendix (except for linguistic reasons), and some matters and terms that scholars specialized in Chinese history consider as basic knowledge are explained somewhat more precisely.

INTRODUCTION

> The military operations by which this success was won, the marches in the desert, the endurance of extremes of heat and cold, the commissariat supplies in foodless and shelterless regions, the transport arrangements in rugged and precipitous lines of communication, deserve the professional study of soldiers and politicians, as showing what China has done in recent times, and what possibly she may yet do.[1]

The Qing dynasty (1644–1911), founded by the Tungusic people of the Manchus, was the last dynastic house ruling over China. Compared to their predecessors, the Ming (1368–1644), the conquest dynasty of the Qing considerably expanded the territory of China proper and created one of the largest states in modern history. The expansion of the Qing and the integration of formerly more or less independent territories has been recently investigated by several authors, like C. Patterson Giersch who studied the integration of the Yunnan frontier region into the empire by a combination of military power and instrumentalization of the indigenous political regimes,[2] Dai Yingcong who argues that the military activities in the southwest and Tibet critically changed the socio-economic landscape of the province of Sichuan,[3] or Peter Perdue in his *China Marches West* who argues that the grip on Central Asia played an important part in world history because China's or Russia's victories in this region "fundamentally transformed the scale of their world" and closed "the great frontier". Soldiers, settlers, missionaries and traders followed each other to the conquered 'New Territories' (*xinjiang*) in Eastern Turkestan, differing in nothing from the Spanish, French or English conquests in the Americas.[4] The important role of merchants for financing and administrating the

---

[1] Richard Temple and Thomas Wade, "Political Lessons of Chinese History", in *Transactions of the Royal Historical Society*, New Series, vol. 1 (1883–1884), p. 228.

[2] C. Patterson Giersch, *Asian Borderlands: The Transformation of Qing China's Yunnan Frontier* (Cambridge etc.: Harvard University Press, 2006).

[3] Dai Yingcong, *The Sichuan Frontier and Tibet: Imperial Strategy in the Early Qing* (Seattle: University of Washington Press, 2009).

[4] Peter C. Perdue, *China Marches West: the Qing Conquest of Central Eurasia* (Cambridge: Harvard University Press, 2005), p. 10.

new territories has likewise been described in James A. Millward's book *Beyond the Pass*.[5]

Peter Perdue has here and there touched upon the issue of organising such large-scale conquest wars,[6] but it remains a point of secondary importance in his book. The present book attempts to fill this gap and delivers a thorough study on how the Qing emperors financed and organised their wars of expansion, especially during the period of time when they had learnt how to do this in the most efficient way. It focuses on one of the most expensive and long-lasting wars in the border region of the province of Sichuan, the second Jinchuan campaign (1771–1776), as a consequence of which Eastern Tibet was incorporated into the Chinese empire.

Logistics in War

Logistics makes up nine tenths of the business of war, says Martin van Creveld,[7] who was one of the first Western scholars systematically studying the art of supplying war. The technical, even mathematical, nature of "providing for the successive arrival of convoys of supplies" has rarely attracted the interest of those exploring the history of warfare.[8] Yet the business of supplying an army deeply influenced the art of warfare and made strategy a mere "appendix of logistics" and service forces more numerous than combat forces.[9]

The present book is about logistics and finance because these two matters cannot be separated. An army cannot do without food, nor are soldiers willing to fight without being paid decently—although highly motivated armies as those of the French Revolutionaries and the more 'democratic' armies of later ages might have done or do so. These two hamstrings of any army are interrelated functionally and organisationally. Supplies often depend on the ability of the state and the highest commanders to raise money. If operation—and even strategy—is wholly subject to logistics, as van Creveld argues, logistics is essentially dependent on cash, leading

---

[5] James A. Millward, *Beyond the Pass: Economy, Ethnicity, and Empire in Qing Central Asia, 1759–1864* (Stanford: Stanford University Press, 1998).

[6] See for instance, Perdue, *China Marches West*, pp. 184, 201, 232.

[7] Martin van Crefeld, *Supplying War: Logistics from Wallenstein to Patton* (Cambridge: Cambridge University Press, 1977), p. 231.

[8] There is a new study on logistics in antiquity by Jonathan P. Roth, *The Logistics of the Roman Army at War (264 BC–AD 235)* (Leiden: Brill, 2012).

[9] Van Crefeld, *Supplying War*, pp. 233–235.

to the necessity to construct a "financial chain of command".[10] Michael Barthorp describes this interrelation as follows:

> In the vital fields of supply and transport, on which the success or failure of a campaign so greatly depended, everything was in the hands of civilian contractors, working to the orders of Marlborough's staff through the agency of Field Commissaries appointed by the Commissioners of Supply and Transport. The efficiency of their arrangements depended on the ready availability of cash, the honesty of the contractors and the resources of the theatre of operations.[11]

It is hardly possible to have armies supplied only by 'living off the country', nor is it feasible that an army lives exclusively on the supplies kept ready in granaries and magazines. In any war, the commanders have to fall back on private suppliers that ship flour, bread and other consumables to the camps and have to be paid for their services. There are many different ways to 'live off the country', as John A. Lynn says,[12] with different grades of arbitrariness and regularity, beginning with plundering and requisition, requiring 'contributions' from towns, to the normal purchase from local markets in enemy territory. Similarly, there are many different methods to raise funds. This can be done by introducing special levies in the own country, by imposing 'war taxes' on adversarial or even neutral communities, or by borrowing money to be paid back later along with a certain interest. Money certainly serves to pay an army, to prevent strikes, desertion or mutiny, but its main purpose is to make sure that the "ambulant city"[13] of tens of thousands of soldiers, servants, sutlers, in earlier ages also the families of the warriors, with tens of thousands of horses or mules and countless wagons and carts, is supplied with food, fodder, equipment and ammunition.

The magnitude of armies between 1600 and 1800 made the business of organising supplies and money a Sisyphean task that was to be shared by various actors. While fundraising was more a business of the war-waging

---

10 Guy Rowlands, *The Financial Decline of a Great Power: War, Influence, and Money in Louis XIV's France* (Oxford: Oxford University Press, 2012), p. 19.

11 Michael Barthorp, *Marlborough's Army, 1702–11* (London: Osprey, 1980), p. 16. The statement refers to the First Duke of Marlborough's (1650–1722) campaigns during the War of Spanish Succession (1701–1713).

12 John A. Lynn, *Feeding Mars: Logistics in Western Warfare from the Middle Ages to the Present* (Boulder: Westview, 1993), p. 16.

13 Van Creveld, *Supplying War*, p. 2.

actor, the state or its "military enterpriser",[14] the organisation of supplies was incumbent on the generals and their staff. The French king in the late seventeenth century therefore created the temporary office of the *intendant général* who was entrusted with the calculation of the amount of consumption, its cost and the cost of transport, and to look after its arrangement. Only in 1783 the Hapsburg monarchy invented a specialized supply corps, and in 1807 Napoleon operated with a train service consisting of fully militarized personnel and equipment.[15]

A careful preparation of supplies is highly important, but not in each case successful, as Napoleon's march on Moscow in 1812 showed. Yet the "tyranny of logistics"[16] at all events had and has a critical influence on the art of warfare. Siege warfare, often called the preferred way of making war in the age of absolutism, was particularly problematic because in this case a large army stayed in one single spot for a long time. Grain reserves and fodder would soon be used up, and the besieging army could not but be supplied from the rear, with the help of an "umbilical cord" to the magazines.[17] On the other hand this type of strategy made it possible to establish regular supply lines, while for a moving army the logistics apparatus has invariably to follow suit. War historians have observed that the standing armies of eighteenth century Europe operated with the help of 'rolling magazines', while Napoleon preferred a kind of 'predatory warfare'. Van Creveld has demonstrated that this concept is incorrect. He divides the history of supplying war according to the main items that the armies needed, namely food and fodder (for men and horses), ammunition (World War I), and fuel (in mechanised warfare).[18] James A. Lynn for his part has shown that it was not so much a revolution in logistics that changed the art of military operation, as van Creveld says, but rather the will to abandon the sterile method of siege warfare.[19]

This theoretical framework provides an important tool to look for the reasons why the emperors of the Qing dynasty were able to establish such a vast empire and to defend it successfully against internal rebellions and external threats. The conquest of Eastern Turkestan, the integration of

---

[14] Larry Neal, *War Finance* (Aldershot: Edward Elgar, 1994), p. xvii, according to the famous books by Fritz Redlich, *The German Military Enterpriser and his Work Force: A Study in European Economic and Social History* (Wiesbaden: Steiner, 1964).

[15] Van Creveld, *Supplying War*, pp. 30, 71.

[16] Van Creveld, *Supplying War*, p. 36.

[17] Van Creveld, *Supplying War*, p. 24.

[18] Van Creveld, *Supplying War*, pp. 232–233.

[19] Lynn, *Feeding Mars*, p. 10.

Tibet, and the cultural and administrative integration of border provinces as Gansu, Yunnan and Taiwan into the Chinese orbit were only possible with the aid of a mighty logistics machine. The last part of this book is dedicated to the question how this engine worked, by whom it was operated, and what fuel kept it going. The answers to these questions demonstrate that the Qing dynasty was able to effectually mobilize the whole government structure and a large part of the population in order to fulfil its ambitious imperial projects in spite of a narrow financial base and a thin bureaucratic network.

War Finance

All governments attempt to wage war as cheaply as possible, yet they often face the predicament of trying to pursue a policy of financing campaigns from the state treasury directly. This was the case under the French Sun King Louis XIV (r. 1643–1715) as well as under his Eastern counterparts, the great Qing emperors of the Kangxi (r. 1662–1722), the Yongzheng (r. 1722–1735) and the Qianlong (r. 1735–1795) reigns.[20] War finance is probably a more attractive theme than the dull technical business of logistics, and therefore worth probing common grounds and differences between various societies and states. The newly emerged East-West debate[21] has not yet dedicated thoughts to war finance, although the fundamental issues of how to pay for war have been, as Larry Neal says, "at the root of macroeconomic policy determination for a very long time".[22]

In Europe the state budgets never sufficed to cover these expenses, so that from the Middle Ages on all governments in the Western world relied on credit, often "borrowing to the limit",[23] mainly from Italian bankers. On the British Islands, the need to finance the war against France led to the creation of the Bank of England in 1694, which was able to provide

[20] Lynn, *Feeding Mars*, p. 138. Dai Yingcong, "Yingyun shengxi: Military Entrepreneurship in the High Qing Period, 1700–1800", in *Late Imperial China* 26/2 (2005), pp. 5–6. According to Chinese custom, the reign mottoes of the emperors are used, and not their personal names, especially for the Qing period. For more details see Appendix 1.

[21] This debate was probably reinitiated by the book of Kenneth Pomeranz, *The Great Divergence: China, Europe, and the Making of the Modern World Economy* (Princeton: Princeton University Press, 2000), but goes back to a book by Eric L. Jones, *The European Miracle: Environments, Economies, and Geopolitics in the History of Europe and Asia* (Cambridge: Cambridge University Press, 1981).

[22] Neal, *War Finance*, p. xi.

[23] Rowlands, *Financial Decline*, p. 72.

the government with cheaper credit than the private market.[24] Other means of expanding the financial scope were introducing regularly collected taxes in money, sale of offices, alienation of future revenues, state-operated lotteries, or exploiting the officers, especially the lower nobility, who privately advanced money for the initial mobilisation, as was the case in France.[25] In both regions of the world, colonels and captains footed the bill for their units and advanced money to the government in urgent cases, hoping (often in vain) to be reimbursed later by the war treasury.[26] In China it was often rather the civil officials working in logistics who had to advance money. The common people could be taxed "to the hilt"[27] not only by extraordinary taxes but also in an indirect way through contractors who took from the local population what the army did not pay to them.[28] The Chinese people was accustomed to be skinned by the local government also in peacetime,[29] a fact well known to and tolerated by the central government, and therefore an opportunity to squeeze money from these officials in turn, when they served as logistics managers in wartimes. Their share to finance the wars of the Qing dynasty was euphemistically termed 'contributions', quite similar to the huge sums of money that the prosperous salt merchants in China 'voluntarily' contributed to the public cause. In Europe the term 'contributions' has a thoroughly different meaning, referring to an explicitly enforced payment, preferably by a community in the territory of the enemy.[30] Any enforced character of the Chinese

---

[24] Michael Godfrey, *A Short Account of the Bank of England* (London: John Whitlock, 1695; republ. in: Michael Collins, ed., *Central Banking in History* [Aldershot: Hants/Brookfield, Vt.: Elgar, 1993], pp. 1–8, also published in Early English Books Online, http://eebo.chadwyck.com/search/full_rec?SOURCE=pgimages.cfg&ACTION=ByID&ID=V36202], p. 2). Godfrey argues: "It will likewise save the Nation 30 or 40000 l. [pounds] *per annum, in remitting the Publick Moneys* [italics in the original]; and if the Bank has been settled in the beginning of this War, *the Kingdom might have saved* already at least *a Million and half, if not Two Millions* by it, besides what will be saved for the future on the Prices of Commodities, Loss on Tallys, Publick Remittances, Excessive Interest, and other Exorbitant Allowances, which *the Publick* has formerly paid for Money, (to say nothing of the Advantage which might have been gained on the Enemy, by having *more early and regular Provisions* made for our Fleets and Armies;) [...]." Compare also the new book by Carl Wennerlind, *Casualties of Credit: The English Financial Revolution, 1620–1720* (Cambridge, Mass.: Harvard University Press, 2011), pp. 109–114.

[25] Neal, *War Finance*, p. xvii. See also Lynn, *Feeding Mars*, pp. 24–27.

[26] Lynn, *Feeding Mars*, p. 138.

[27] Rowlands, *Financial Decline*, p. 57.

[28] Van Creveld, *Supplying War*, p. 38.

[29] Ch'ü Tung-tsu, *Local Government in China under the Ch'ing* (Cambridge, Mass.: Harvard University Press, 1962), pp. 22–32.

[30] Redlich, *Military Enterpriser*. Lynn, *Feeding Mars*, p. 145, demonstrated that under Louis XIV the share of contributions to war expenditure was relatively low. In 1692

contributions was furthermore somewhat attenuated by the bestowal of a brevet title of a state office on the contributor, a practice that became a regular way of filling the state treasury also in peacetime from the mid-nineteenth century on, supplementary to the newly introduced *likin* (*lijin*) transit tax.[31] 'Contributions' in the Chinese context were therefore similar to the sale of offices that was widely practised in absolutist France.[32]

In the age of absolutism warfare was a kind of sports[33] of the ruling elite, which required ever increasing funds. In Europe, heavier taxation and reliance upon credits was often the only means to avoid conflicts with parliaments or estates of the realm. No such conflicts between political actors arose in China, but the concept of government debt was totally unknown. Consequently, the Qing state waged war with a rather slender budget and therefore cut cost wherever possible. The attempts at curtailing expenditure culminated in the creation of a War Supply and Expenditure Code (*Junxu zeli*)[34] in 1785 that made supply and logistics, and the expenditure for it, a bureaucratic nightmare with a labyrinthine auditing process that sometimes ended in the emperor's desperate cancellation of private debts accumulated by logistics officials. This limited budget forced the Qing government to walk at the edge of abysmal depths, confronted

---

contributions made out only 4 per cent of the budget, and in other cases 18 per cent at the most.

[31] See Elisabeth Kaske, "The Price of an Office: Venality, the Individual and the State in 19th Century China", in *Metals, Monies, and Markets in Early Modern Societies: East Asian and Global Perspectives*, ed. by Thomas Hirzel and Nanny Kim (Berlin: LIT Verlag, 2008), pp. 281–308. Elisabeth Kaske, "Fund-Raising Wars: Office Selling and Interprovincial Finance in Nineteenth Century China", in *Harvard Journal of Asiatic Studies* 71/1 (2011), pp. 69–141. Edwin George Beal Jr., *The Origin of Likin, 1853–1864* (Harvard: Harvard University Press, 1958).

[32] William Doyle, *Venality: The Sale of Offices in Eighteenth-Century France* (Oxford: Oxford University Press, 1996).

[33] William McNeill, *The Pursuit of Power: Technology, Armed Force and Society since A.D. 1000* (Chicago: University of Chicago, 1982), p. 161.

[34] *Junxu zeli* (Xuxiu siku quanshu, vol. 857). It is composed of three parts, according to bureaucratic responsibility: the *Hubu junxu zeli* (Regulations of the Ministry of Revenue for war supply and expenditure), *Bingbu junxu zeli* (Regulations of the Ministry of War for war supply and expenditure), and *Gongbu junxu zeli* (Regulations of the Ministry of Works for war supply and expenditure). The term *junxu* means, according to the dictionary *Hanyu da cidian*, ed. by Luo Zhufeng (Shanghai: Cishu chubanshe, 1986–1993), vol. 3, p. 1212, 'all food, clothes, tools, implements and all other goods and materials', but in all Qing period sources I came to use, *junxu* has definitely also—and often exclusively—the meaning of 'money, funds'. The *Junxu zeli* is mainly concerned with cutting expenses by fixing ceiling amounts of allowances (monetary and non-monetary, like horses or manservants), for which reason a satisfactory translation can only be 'supply and expenditure code', although, for the purpose of convenience, I may also call it 'war expenditure code' from time to time.

with the need to nourish the military hereditary class as the conquest elite upon whose support the Manchu dynasty relied.[35] Only a war offered the professional military in China who were paid relatively modest salaries in peacetime an opportunity to become wealthy by profiting from stipends and rewards.[36] To a certain extent, warfare was necessary in order to secure the soldiers a certain standard of living. Soldiers as a hereditary class were in China therefore also far less prone to desertion than their counterparts in the West, where armies consisted of the "scum of every country" and the "refuse of mankind"[37] that displayed no devotion to their employer and fought for whoever was able to pay in time.

Last, but not least, it is important to note that in China as well as in Europe logistics officers heavily relied on private entrepreneurs and their financial potency to feed and equip the army during campaigns. Merchants played a vital role in providing food and transport, not only on a large scale, but even down to individuals owning a waggon, cart or mule. "Bureaucratisation of violence"[38] therefore does not only mean more red tape, but also a more civilised way of interaction between government and private actors. In his recent book Guy Rowlands even argues that "a set of individuals and loose groups", which he calls "military-industrial complex", did not only support the war effort of the king of France, but also "augmented the state's financial distress".[39] Such a great influence of private actors on the state budget cannot be observed in China, but it is a known fact that local merchants made great profits from supplying the army.[40]

The crucial part—or 'root', as the Chinese would say—that money plays for belligerent activities of all states in the world is treated at the beginning of this book with the themes of expenditure and finance, while

---

35 Mark Elliott, *The Manchu Way: The Eight Banners and Ethnic Identity in Later Imperial China* (Stanford: Stanford University Press, 2001), pp. 306–31.

36 Dai Yingcong, "Military Finance of the High Qing Period: An Overview", in *Military Culture in Imperial China*, ed. by Nicola di Cosmo (Cambridge: Harvard University Press, 2009), pp. 304, 309.

37 Quotation of Campbell Dalrymple's booklet *A Military Essay: Containing Reflections on the Raising, Arming, Cloathing, and Discipline of the British Infantry and Cavarly* (London: Wilson 1761; also to be found on http://books.google.de/books/about/A_military_essay.html?id=xyNEAAAAYAAJ&redir_esc=y.), p. 8. Thanks are due to Peter Way, University of Windsor, Ontario, for pointing out this essay to me.

38 McNeill, *Pursuit of Power*, p. 144.

39 Rowlands, *Financial Decline*, p. 14.

40 Perdue, *China Marches West*, p. 201.

logistics—the 'branches' in Chinese thinking—is dealt with in the last third.

The Relevance of Case Studies

The study of individual campaigns like the Duke of Marlborough's march to Blenheim (Höchstädt) on the Danube in 1704, Napoleon's campaign against Austria in 1805 or his attempt to force the Tsar to his knees in 1812, or, as in this book, that of the second Jinchuan war, helps to find out why commanders sometimes achieved victory, and why they failed in other cases. Case studies help to define parameters like instruments, actors and conditions, but also individual decisions and political determinations. This is especially interesting and all the more visible in an intercultural comparison. The present volume is therefore also based on an in-depth study of one single campaign, enriched by observations from other Qing wars between 1650 and 1800.

China was no less belligerent than the European states in the seventeenth and eighteenth centuries. Of the many military campaigns that are frequently mentioned throughout this book, the second Jinchuan war is most excellently documented, especially regarding logistics and finance. It is therefore an ideal case to serve as a paradigm for how the business of supplying armies and providing funds worked in late imperial China. It is even all the more interesting as it was the longest-lasting and also the most expensive single campaign before 1800. The excessive cost for transport was admittedly also caused by the adverse topographical conditions in high mountain ranges, yet seen from the organisational side, this war proves a perfect example of how various actors worked together, and what sources were tapped to comply with the Confucian image of a benevolent and even-handed government. The Jinchuan war was also one of the last great wars of expansion before the Qing empire reached its limits, turned inward and was bogged in the mire of social and economic problems.

The year 1776 can therefore be seen as a kind of watershed between a long phase of activity and prosperity and a new era of "decline", "inefficiency", "corruption", "degradation" and "irresponsibility".[41] On the other side of the world, the same year inaugurated the dawn of a very different

[41] Keywords according to Immanuel C. Y. Hsü, *The Rise of Modern China* (New York/Oxford: Oxford University Press, 2000[7]), pp. 123–127.

kind of age, and might therefore symbolize a kind of "great divergence" between East and West.

China had, of course, a distinct "culture of war",[42] in which soldiers occupied a privileged position in society and the population was permanently aware of wars which were brought to their attention, by commemorations in various forms. Also the concept of building an empire and uniting the diverse peoples living within its borders was generally known. Yet there are some other differences in the method and instruments of warfare. John A. Lynn especially stressed the importance of naval warfare in Europe which played an enormous role as early as the sixteenth century and was a crucial factor for the overseas expansion of European commerce and politics.[43] Ships (and also boats, for transport on inland waterways) have the advantage that they can shift huge masses over wide distances at a low price. Warships were furthermore equipped with so much firepower that could not easily be summoned by land armies. Naval warfare—apart from much higher cost than land-based warfare—demands a much different kind of logistics because the crews of ships have to be supplied from the onset and with careful planning in advance. Once out of port there is no opportunity to rely on granaries, warehouses on wheels, markets, or peasants that can be bought or stolen from. In late imperial China, naval warfare had a wholly subordinated role in the military system. Only a few campaigns involved naval transport at all, and even fewer naval battles. Siege warfare, which dominated the art of war in Europe likewise played a minor role in China, hand in hand with the minor importance of artillery. Nicola di Cosmo has shown that artillery played a certain role in the Ming-Qing transition,[44] but not such a dominant one as it did, for instance, in the Ottoman empire whose armies used monstrous bombards to smash the city walls of their opponents, and whose firearms technique was on a par with that of Christian Europe, at least until 1700.[45] The Qing emperors

---

42 Joanna Waley-Cohen, *The Culture of War in China: Empire and the Military under the Qing Dynasty* (London, New York: Tauris, 2006). Nicola di Cosmo, *Military Culture in Imperial China* (Cambridge: Harvard University Press, 2009).

43 Lynn, *Feeding Mars*, pp. 13–14. Concerning the importance of ships in European overseas expansion, see the classical study of Carlo M. Cipolla, *Guns and Sails in the Early Phase of European Expansion, 1400–1700* (London: Collins, 1965).

44 Nicola di Cosmo, "Did Guns Matter? Firearms and the Qing Formation", in *The Qing Formation in World-Historical Time*, ed. by Lynn Struve (Cambridge, Mass: Harvard University Press, 2004), pp. 121–166.

45 See Gabor Ágoston, *Guns for the Sultan: Military Power and the Weapons Industry in the Ottoman Empire* (New York: Cambridge University Press, 2005), pp. 9, 195–198. Rhoads

nevertheless continuously stressed the advantage of cannons and howitzers in warfare and tried to obtain pieces of better quality.[46]

In spite of all these important differences there are surprisingly many basic similarities not only in war finance, as seen above, but also in the issue of logistics in China and in Europe between 1600 and 1800. Horses and fodder were the most critical issues in the pre-mechanised age. Often enough, half of the transport, riding and draft animals died of exhaustion and starvation. Hungry armies were during the march often divided into several independently operating contingents (the "invention of the division"[47]) that crossed separate regions in order to avoid depleting local flour or grain reserves, stocks of animals, carts for transport, and manpower for serving the army. Waterway transport was chosen, wherever possible, for bulky and heavy goods, but also for troops, saving the energy that was otherwise consumed during enforced marches. Sparsely inhabited regions were crossed as quickly as possible. In any case, local officials were notified in advance of the date of the arrival of troops and the quantities of food they were to be provided with. Governments and their generals all over the world made use of the civilian bureaucracy to serve the needs of the military apparatus. The unattractive logistics machine was frequently operated by discharged officers, punished in this way for their incompetence in the field.[48] The Chinese government also made use of unpaid appointees for a government position to take over tasks in the logistics apparatus, and so saved a lot of money. There were, on the other hand, also instances that men who worked in the grain supply would receive equal military honours with soldiers who fought in battle.[49]

Comparative research can thus not only be an instrument to highlight differences in the social, political and economic development between European states and China, but also to underline common ground in the *conditio humana*, as has been done, for instance, in a recent publication by the School of Humanities and Social Science, Hong Kong, on human

---

Murphey, *Ottoman Warfare, 1500–1700* (New Brunswick, N.J.: Rutgers University Press, 1999).

46 See, for instance, Perdue, *China Marches West*, pp. 175, 184, 305, or Joanna Waley-Cohen, "God and Guns in Eighteenth-Century China: Jesuit Missionaries and the Military Campaigns of the Qianlong Emperor (1736–1795)", in *Contact between Cultures*, vol. 4, *Eastern Asia: History and Social Sciences*, ed. by Bernard H. K. Luk (Lewiston: Mellen, 1992), pp. 94–99.

47 McNeill, *Pursuit of Power*, p. 162.

48 Van Creveld, *Supplying War*, p. 43.

49 Perdue, *China Marches West*, p. 183.

motivation, the reproduction regime, and population behaviour in Europe and Asia,[50] or by the ethnologist Christoph Antweiler who operates with a concept of inclusive humanism instead of trying to find out differences between peoples.[51] On the other hand, some crucial differences between the two cultural realms of the 'Western world' and China may perpetuate and still be visible in the present age, like the tendency to finance consumption (or war) with the help of credit, against a tradition of high saving rates, both in the private and the public spheres. The seeming ease of fundraising and ostensible abundance of state reserves in China, today as in the eighteenth century, is in sharp contrast to the permanent indebtedness of European states, under Louis XIV as well as nowadays. Some observations in the issue of fundraising might even challenge time-tested perceptions of forms of government—the word 'absolutism' (in the sense of 'not being accountable') probably applies more to the imperial Chinese state with its 'bureaucratic absolutism' than to France around 1700.[52]

The Jinchuan Issue

The region of Jinchuan was located in the western fringes of the modern province of Sichuan. It was originally not part of that province, but was ruled by native chieftains or kinglets (*tusi* 'local leaders')[53] that were by the Chinese emperors granted the title of imperial 'pacification

---

[50] Noriko O. Tsuya et al., *Prudence and Pressure: Reproduction and Human Agency in Europe and Asia, 1700–1900* (Cambridge, Mass.: MIT, 2010).

[51] Christoph Antweiler, *Inclusive Humanism. Anthropological Basics for a Realistic Cosmopolitanism* (Göttingen: V&R Unipress and Taipei: National Taiwan University Press, 2012).

[52] The term 'bureaucratic absolutism' has been coined by Thomas Ertman, *Birth of the Leviathan: Building States and Regimes in Medieval and Early Modern Europe* (Cambridge: Cambridge University Press, 1997), pp. 6–10, 224–263, to describe the post-reform state of Prussia in the early nineteenth century, but it is occasionally also used to describe the political system of late imperial China, e.g. Charles O. Hucker, *China's Imperial Past: An Introduction to Chinese History and Culture* (Stanford: Stanford University Press, 1975), pp. 303–307, or Kai Vogelsang, *Geschichte Chinas* (Stuttgart: Reclam, 2012), p. 615.

[53] The terms *tusi* is used as a general term for native chieftains especially of peoples in the southwestern provinces of Sichuan, Yunnan, Hunan, Guizhou and Guangxi. The office of *tusi* as a native ruler endowed with an official seal by the imperial court was invented during the Yuan period (1279–1368) but it was only institutionalized during the Ming period. Charles O. Hucker, *A Dictionary of Official Titles in Imperial China* (Stanford: Stanford University Press, 1985), No. 7355. It was especially during the Qing that these chieftain-led regions were transformed into 'normal' prefectures (a process called *gaitu guiliu* 'transform native rule into the common pattern').

PROVINCES, REGIONS
■ main prefectures, important towns
native kingdoms

Map created by the author, based on Tan Qixiang, *Zhongguo lishi ditu ji*, vol. 8, *Qing shiqi* (Beijing: Ditu chubanshe, 1996), pp. 39–40.

Map 1. The Province of Sichuan (ca. 1750) and the Native Kingdoms in Eastern Tibet (Kham).

commissioner' (*xuanweishi* or *anfushi*)[54] since at least the Ming period and were so part of the system of indirect rule (*jimi* 'slackening reins') over border regions.[55]

54 Hucker, *Dictionary*, Nos. 17, 2682, translates these two titles in the same way, although a *xuanweishi* was of official rank 3b, and an *anfushi* of rank 5b. One might therefore speak of pacification commissioners of 1st and 2nd class respectively. Hippolit Semenovich Brunnert and V. V. Hagelstrom, *Present Day Political Organization of China* (Shanghai: Kelly and Walsh, 1912), p. 439 (No. 861A), do not translate these titles, nor does William F. Mayers, *The Chinese Government: A Manual of Chinese Titles* [...] (London: Kelly and Walsh, 1897), p. 47 (Nos. 332, 334).

55 The *jimi* system had been introduced during the Tang period (618–907) for the indirect administration of territories of subjugated non-Chinese polities or tribes, mainly in the protectorate of the Pacified West (modern Xinjiang and beyond), and also in the southwestern region (modern Yunnan). The local chieftains were granted a hereditary rank of protector, area commander-in-chief, or prefect, and often concurrently bestowed a title of nobility. An official seal handed over by the Tang court was their most important symbol of being part of the Chinese empire, at least nominally.

The conquest of the Jinchuan region was part of the large-scale colonial projects of the Qing emperors that integrated the regions of Eastern Turkestan (Xinjiang), Amdo (Xining, Kokonor, Qinghai) and Tibet (Xizang) into the Chinese empire. The region between Tibet proper and the province of Sichuan, Eastern Tibet or Kham (Khams), was politically and culturally distinct from central Tibet and was only seized by the Qing in several steps, the main reason for the occupation being the unruly character of the local chieftains whose belligerent activities endangered the peaceful traffic between Sichuan and central Tibet.

Jinchuan (in the native language rGyal-rong) actually consisted of two kingdoms, namely Greater Jinchuan (Chu-chen or Rab-brtan, modern Jinchuan) and Lesser Jinchuan (bTsan-lha, modern Xiaojin). The Chinese names for these kingdoms are derived from those of two rivers, the Greater and the Lesser Jinchuan River that are tributaries to the Dadu River. They were surrounded by a dozen other small kingdoms whose rulers were for the most part willing to serve as allies to the Qing government. Part of them, like Dartsedo (Dajianlu, modern Kangding), Djandui (Zhandui, Garzê, modern Xinlong) or Dzagu (Zagunao, modern Lixian), had been submitted by Qing troops some decades earlier. The king of Greater Jinchuan occupied an important religious position as a kind of lord-lama (in the local language *slo-χpɯn*,[56] in Chinese rendered as *shaloben*, in Manchu as *solobun*), secular and religious leader of his people.

The land of Jinchuan is still a quite inaccessible region, whose topography is characterized by precipitous mountains with steep slopes and deep gorges. The extreme character of the topography can be explained by the great incline between the Tibetan highland and the Sichuan Basin which makes creeks and rivers cut in deeply into the geological strata. Even in the valleys of the Greater and Lesser Jinchuan Rivers there was not much space for agriculture. The climate with its great number of rainy days has never been very friendly and it is reported that even in the summer months snow was no uncommon phenomenon.[57] Jinchuan was only accessible via a handful of mountain passes of which some have a height of up to 4,000 metres or more. Rivers and creeks were often simply impassable, and only a few bridges allowed to safely get across gorges or canyons. The natives normally used cow hide boats in order to cross the

---

[56] Guillaume Jacques, *Phonologie et morphologie du Japhug (rGyalrong)* (Ph. Diss.: Paris, 2004), p. 110.

[57] Li Xinheng, *Jinchuan suoji* (Congchu jicheng, vol. 3199), juan 2, pp. 12–13.

Source: *Huang-Qing zhigong tu,* juan 6, fol. 25.

Figure 1. Man and woman from Lesser Jinchuan.

rivers, but there were also some places where makeshift bridges held by bamboo ropes crossed the two Jinchuan rivers.[58]

A first campaign to bring order into the quarrels among the mountain kings of Kham was undertaken between 1747 and 1749. This was the so-called first Jinchuan war that was triggered when Greater Jinchuan occupied Lesser Jinchuan and made raids on the villages of other local kingdoms. The campaign, led in order to restore the order that normative control was not able to achieve,[59] ended in a truce that the young Qianlong emperor at the time hailed as his first military victory. In fact the emperor was galled by the extremely poor performance of his highest generals, two of whom, governor-general Zhang Guangsi and Grand Minister Neqin, engaged in a private quarrel instead of fighting the enemy. They were also at odds about the strategy of fighting the infamous war-towers (*diao*) of the natives, some of which can still be seen today in the region. Usually houses were made of crude stones and therefore could not only withstand the harsh winter breezes from the glaciers but also resist any attack by marauders from the neighbourhood. Because the inhabitants

[58] *Jinchuan suoji,* juan 2, p. 16.

[59] Compare R. Kent Guy, *Qing Governors and their Provinces: The Evolution of Territorial Administration in China, 1644–1796* (Seattle: University of Washington Press, 2010), p. 209.

Source: *Huang-Qing zhigong tu*, juan 6, fol. 27.

Figure 2. Man and woman from Greater Jinchuan.

of Jinchuan were frequently afflicted with the shortness of food they resorted to a lifestyle that was characterized by simple robbery or banditry.[60] When any village in Jinchuan suffered from hunger, it was normal for people to band together in order to take food from a neighbouring village or from a village in the territory of another native king. Stealing cattle or taking hostages in order to obtain ransom in the shape of food had therefore become an integral part of the hard life of the population of Jinchuan. Their war-towers were an ideal protection for each house in case of a siege: They were easy to build and easy to defend even against a large number of attackers. It was especially the seats of village heads (*guanzhai*) that were protected by many war-towers grouped in several circles around the central castle and the glacis.

Zhang Guangsi reported to the throne and explained the overall situation in the war area:[61] The territory was characterized by precipitous mountains reaching up to the sky, which made any move to advance extremely difficult. Furthermore, the rebels had erected war-towers

[60] *Jinchuan dang*, ed. by Guoli gugong bowuyuan diancang zhuan dang'an ji fanglüe congbian (Taibei: Shenxiangting qiye, 2007), 37/IV/00458 (QL 37/12/protocol).

[61] *Pingding Jinchuan fanglüe* (Siku quanshu edition), juan 3, fol. 18a–21a (QL 12/9/yiwei).

wherever possible. These towers were constructed from stones piled up like a city wall. The highest of them had many stories and were as tall as 16 *zhang* (about 51 metres). On all sides the towers were equipped with embrasures that made it possible to observe the surroundings without being seen by attackers. The defenders meanwhile were able to conveniently shoot down at the Qing troops who had no chance to see the enemy. In some places the towers did not stand isolated but were clustered together in a kind of fortification where each tower could protect the neighbouring one. Around them walls and moats made any approach impossible. Common methods for conquering a fortress, like sapping the walls or digging tunnels in which mines could be placed, or to erect walls for a battery, from which the fortifications could be bombarded, or to surround the castle to starve the enemy, could not be employed in Jinchuan because the rebels found it easy to counter such methods. The experience during the rebellion in Djandui (1745) had taught the Jinchuan rebels to be prepared: they either dug wide and deep moats that could not be crossed, or stored enough water and eatables inside their castles, or erected walls and other fortifications around the castle and the towers. Of immense help to the rebels were the physical conditions of the territory: cliffs and steep slopes made attacks virtually impossible. Even the cannons and howitzers could not be used because there was often no place to erect a battery platform. In order to destroy the war-towers the Qing army needed heavy artillery, but many of them were erected in places the cannons could not even be taken to, neither by mules nor by manpower. Smaller cannons had almost no effect on the construction of the war-towers, and the rebels could easily repair damage caused by them. Another method to destroy the war-towers was the old method of burning them down. The soldiers cut down large trees and transported the wood to the foot of the towers. During that work they had to be protected by wooden shelters because the rebels could easily fire at the defenceless soldiers transporting the fuel. Fire-tipped arrows to burn down barns and granaries, a common method when assailing a village or a city, were also of no use, because all buildings were made of stone and the food was stored deep inside the towers. Another reason why the method of burning down the castles was often unfeasible was the regular rainfall during the whole year, and in winter the snow made it often impossible to move forward at all. The tactics of the native auxiliaries who were accustomed to conquer such towers by climbing up from the outside and fighting against the defenders from the top of the building could not be applied either, as there the rebels could easily fire at the invaders.

In spite of experiences from several campaigns in the Tibetan highland a generation earlier, the rough territory of Jinchuan posed difficulties for the Qing troops that were more accustomed to fighting in flat territory where cavalry units could charge the enemy. Native allies could never be trusted. Indeed did spies have access to secret information from headquarters and transmitted it to the rebels who took measures to counter the plans of Zhang Guangsi. Charged with high treason, incompetence and waste of state funds, he was executed, while Neqin was allowed to kill himself with his grandfather's sword.[62] In early 1749 they were replaced by Grand Minister Fuheng and General Yue Zhongqi who brought the campaign to a quick end. It had cost about 7 million silver *liang*.[63]

The continuing quarrels between the native kings in the region resulted in a second, protracted campaign in the high mountains of Jinchuan which cost the Qing government no less than 62 million silver *liang*, or the whole government revenue of one year.[64] It was the longest, most intensive and most expensive campaign that the Qing dynasty had fought in the eighteenth century. It involved 120,000 troops, about 400,000 civilian labourers and a large amount of personnel from the civilian bureaucracy throughout the empire that operated the logistics apparatus.[65] It showed that the officers as well as the civilian officials immensely profited from funds and materiel poured into the war chest, and it became necessary to finally fix rules for the accounting of war expenditure. Fortunately a sufficient number of documents from different collections of source material supply information on what practices for logistics, finance and settlement of accounts were applied during this crucial war as part of the

[62] Wei Yuan, *Shengwuji* (Sibu beiyao edition), juan 7, *Qianlong chu ding Jinchuan tusi ji*. See the biographies of both persons in Ke Shaomin, *Qingshigao* (Beijing: Zhonghua shuju, 1977), juan 297 and 301, or *Qingshi liezhuan*, ed. by Wang Zhonghan (Beijing: Zhonghua shuju, 1987), juan 17 and 22, or Arthur W. Hummel, *Eminent Chinese of the Ch'ing Period* (Washington: US Government Printing Office, 1944), pp. 43–45.

[63] *Liang*, in the West also known by the Malaiian name *tael*, is a unit of weight of about 37.7 g. It was also the highest denomination for the silver currency (which, as commodity silver, had to be assessed for quality and weighed on exchange), the sub-units being 'mace' (Chinese *qian*), 'candareen' (*fen*) and 'cash' (*li*). As a currency unit it also served as the standard monetary unit for official accounting.

[64] Chen Feng, *Qingdai junfei yanjiu* (Wuchang: Wuhan daxue chubanshe, 1992), pp. 261–262; and Faššan (Fashishan), *Taolu zalu* (Beijing: Zhonghua shuju, 1959), juan 1, pp. 24–26.

[65] Dai Yingcong, "The Qing State, Merchants, and the Military Labor Force in the Jinchuan Campaigns", in *Late Imperial China* 22/2 (2001), pp. 35–90.

Qianlong emperor's "imperial project".[66] This circumstance makes it all the more worthwhile to undertake an in-depth study of this war and its organization.

We owe the first study on the Jinchuan wars to Erich Haenisch,[67] who analyzed the most important historiographic sources dealing with them. He used the report from the mid-nineteenth century book *Shengwuji*, which is a glorification of the Qing military. Roger Greatrex has supplied a thorough analysis of the first Jinchuan war and its background.[68] In his work he shows how the habitual raids of the local people became rampant in the course of time, which finally triggered the decision of the Qing government to stop the permanent quarrels in the border areas of Sichuan once and for all, and in a brutal way. Thus his study complements the earlier works of Patrick Mansier and Dan Martin[69] who particularly stress the religious background of the second Jinchuan war and show how the Qianlong emperor as patron of the 'Yellow Hat' dGe-lugs-pa (Gelugpa) school was urged to put an end to the rule of the native kings of Jinchuan who for their part protected the 'heretic' Bön (Bon) school.

The second Jinchuan war erupted in the summer of 1771 after the lord-lamas of Greater and Lesser Jinchuan, Sonom and Senggesang, joined forces to form a powerful block that threatened the peace of the whole region. The *casus belli* was thus similar to the activities of Galdan, Tsewang Rabdan and Amursana, leaders of the Dzungars, who had threatened the stability of the Western Territories and even that of Tibet in

---

66 Mark C. Elliott and Peter Stearns, *Emperor Qianlong: Son of Heaven, Man of the World* (Upper Saddle River: Longman, 2009), pp. 86–89. Pamela Kyle Crossley, *The Manchus* (Cambridge, Mass.: Blackwell, 1997), pp. 112–117.

67 Ernst Haenisch, "Das Goldstromland im chinesisch-tibetischen Grenzgebiete, nach dem großem Kriegswerk vom Jahre 1781 dargestellt", in *Southern Tibet*, vol. IX, ed. Sven Hedin and Albert Herrmann (Stockholm: Lithographic Institute of the General Staff [...], 1922), pp. 67–132. Ernst Haenisch, "Die Eroberung des Goldstromlandes in Ost-Tibet als Beitrag zur chinesischen Kolonialgeschichte des 18. Jahrhunderts", in *Asia Maior* X (1935), pp. 17–21.

68 Roger Greatrex, "A Brief Introduction to the First Jinchuan War (1747–1749)", in *Tibetan Studies: Proceedings of the 6th Seminar of the International Association for Tibetan Studies, Fagernes 1992*, ed. by Per Kværne (Oslo: The Institute for Comparative Research in Human Culture, 1994), pp. 247–263.

69 Patrick Mansier, "La guerre du Jinchuan (rGyal-rong): Son contexte politico-religieux", in *Tibet: Civilisation et Société, Colloque organisé par la Fondation Singer-Polignac à Paris, les 27, 28, 29 avril 1987*, ed. by Fondation Signer-Polignac (Paris: Éditions de la Maison des sciences de l'homme, 1990), pp. 125–142. Dan Martin, "Bonpo Canons and Jesuit Cannons: On Sectarian Factors Involved in the Ch'ien-lung Emperor's Second Goldstream Expedition of 1771–1776", in *The Tibet Journal* 15/2 (1990), pp. 3–28.

earlier decades.[70] Governor-general Agui (1717–1797), who took over the command of the campaign, rose to one of the most prominent figures in Qing politics. He was able to liberate Lesser Jinchuan in January 1773, but only after the loss of the lives of several high commanders. One of them, former governor-general A'rtai, served in logistics as a punishment and was finally allowed to commit suicide because of blatant embezzlement.[71] From Lesser Jinchuan, the troops were to cross the mountains and to engage the main castles of Greater Jinchuan, Le'uwei (modern Jinchuan) and Garai (modern Anning), from several sides. Like during the first Jinchuan war, the Qing troops only advanced very slowly. In July 1773 a terrible disaster shook the whole imperial army: spies had been able to sneak in general Wenfu's quarter in Mugom and had helped to prepare a large-scale assault on the camp and the supply lines between Lesser Jinchuan and the front. Thousands of troops and labourers were killed or captured. Wenfu died, too, luckily, for he would without doubt have been executed for his failure to secure the hinterland and the camp itself.

The whole Jinchuan army had to be reorganized and was reinforced after this setback. An infusion of fresh troops, including elite Banner troops from Beijing and the northeastern provinces, was necessary to stabilize the front and to make a new assault. The Mugom incident was the decisive factor for the emperor's decision to wipe out Jinchuan once and for all. For this purpose, he was not closefisted but allowed the Ministry of Revenue to transfer immense amounts of money to the war chest in Sichuan. In a combination of scorched-earth tactics and heavy barrage of its main castles Jinchuan was purged of rebellious elements. The conquest of the two main lamasery castles was only possible by employing a Jesuit geodesist, Felix da Rocha (1731–1781), who instructed the artillerymen in the correct use of the howitzers that crushed the fortification walls. The last stronghold was assailed and conquered in January 1776. The main scoundrel, lord-lama Sonom of Greater Jinchuan, and his family, were transported to the capital where they were presented to the emperor and then cruelly executed. The rest of his people was enslaved and presented to the allied local kings. Also in this respect, the treatment

70 Perdue, *China Marches West*, pp. 227–232.

71 *Shengwuji*, juan 7, fol. 20b. *Qingguoshi*, ed. by Wu Ge (Beijing: Zhonghua shuju, 1993), *Dachen zhengbian*, juan 147, p. 925. For a comparison with Neqin, see *Jinchuan dang* 36/IV/00141 (QL 36/10/9), 00147 (QL 36/10/14), and 00167 (QL 36/11/1). His biography is to be found in *Qingshigao*, juan 326.

of stubborn enemies was similar to the genocide of the Dzungars in 1757.[72] After the victory over Sonom in 1776 Jinchuan was transformed into a directly administered sub-prefecture (*zhili ting*) and settled by Chinese immigrants and military colonists.

The subjugation of the Tibetan people of Jinchuan was practically the last colonial war of the Qing dynasty. The victory of the Qing army coloured a 'white spot' on the map of the empire. It initiated the final transformation of all former *tusi* territories into regular administration units and their integration into a seemingly uniform empire.[73] The Qing empire had by 1800 become a modern territorial state with clearly defined borders and direct administrative access on all regions, at least in the form of a Grand Minister Superintendent (*banshi dachen*) in Tibet, Xining, Khotan, Yarkant, Uš, Aksu, Kuča, and other polities in Eastern Turkestan. Subsequent martial actions were therefore exclusively suppressions of rebellions within Chinese territory and at its borders. The only exceptions were the 1788 campaign to ensure the survival of a unified government in Annam, one of the traditional vassal states of China, and the campaigns against Jahāngīr (Chinese rendering Zhang Ge'r) in Turkestan between 1820 and 1830. In these cases it can be seen that a territorial state is not necessarily a nation state acknowledging the equal rights of other nation states. Nor does a territorial state consist of homogeneous units: Ethnic conflicts rumbled on and repeatedly erupted in rebellions against the Han Chinese immigrants and the Qing occupation troops. Jinchuan was indeed pacified once and forever: Its native inhabitants are not granted the status of a National Minority today.

Findings and Relevance of the Topic

The many wars that the Manchu emperors fought were—at least rhetorically—financed exclusively from the state treasury. A substantial part was transferred not from the treasury of the Ministry of Revenue but directly from various provinces to the war-waging province whose governor-general was responsible for the organization of the campaign. During the high

---

[72] About the extirpation of the Dzungars, see Perdue, *China Marches West*, pp. 284–286. According to Perdue, the nineteenth century historian Wei Yuan estimated that 30 per cent of the male population was massacred, and their wives and children forced into slavery. A great part of the rest died of smallpox, a disease to which also many Manchus succumbed in earlier times. See Wei Yuan, *Shengwuji*, juan 4, fol. 8b.

[73] Eastern Turkestan was transformed into the province of Xinjiang in 1884.

Qing period (Yongzheng and Qianlong reigns) about one tenth of the war expenditure, with increasing tendency, was raised as so-called contributions (*juan*) from the prosperous salt merchant associations that had been subsidized by the government in earlier decades. Another source of contributions were private persons, especially state officials who thus purchased brevet titles of an office that gave them a higher reputation. Direct tax increases to finance campaigns were officially inacceptable, yet in some cases, local surcharges were levied in order to finance the supply apparatus. The instrument of credit, be it short-term or long-term, was entirely unknown in traditional China. The government was even able to act as a lender, not as a borrower. It pre-financed the contributions of merchants, the debts accumulated by state officials who spent more money for supplies than officially allowed and even waived taxes in regions obliged to provide military labour for logistics. In spite of deliberately frozen land tax quotas the Qing finances were thought to rest on such solid foundations that the state could easily not only afford to maintain a standing army of nearly a million and to feed their families, but also pay for wars out of petty cash. These expensive wars were fought during a period of "fiscal excellence" partly as a result of huge amounts of silver inundating China through foreign trade.[74]

The sudden financial crash during the Jiaqing reign therefore seems quite a surprising matter on that background. Yet when looking closer at jurisdictional regulations it becomes apparent that the state had always been highly interested in curtailing expenditure. For each war, a procedures code was created in order to check facilities, performance and disbursement in the logistics field. While these procedures differed from war zone to war zone, the nation-wide "War Supply and Expenditure Code" (*Junxu zeli*) was created in 1785 with the intention to limit expenses. Overspending had to be reimbursed by the officials working in logistics. Many of them were even not regularly employed, but served unpaid on probation or because they were on a charge. Soldiers were supported in a hereditary military system, yet the government did neither finance their uniforms nor their weaponry. The largest amount of war expenditure was not for armament or salaries, but for food and transport. In particular the prices of the two items of rice, as a staple food, and transport services

[74] Alexander Woodside, "The Ch'ien-lung Reign", in *Cambridge History of China*, vol. 9, *The Ch'ing Empire to 1800*, part 1, ed. by Willard J. Peterson (Cambridge: Cambridge University Press, 2002), pp. 270–271.

could not be controlled by the government, regardless of any jurisdictional regulations. Rewards was a further item whose cost could not be curtailed by any regulations. While it was possible to fix the amount of rice given to a man and the number of horses an officer could use, it was unfeasible to restrict the amounts of brocades or full monthly surplus stipends granted as remunerations for heroic conduct in battle.

Compared to the West, but also to the Ottoman empire,[75] war in late imperial China was fought at a relatively low level of cost, even in a phase of fiscal prosperity: land warfare did not need high investments in shipbuilding, nor were large fortifications to be constructed. Firearms were of a relatively modest technical level and therefore cheap: matchlock muskets were used until far into the nineteenth century, and guns were frequently acquired as disused objects from the Portuguese and often simply copied. These are further reasons why the Qing emperors could finance their wars to a large extent from the running budget. Their way of organizing supplies was also very efficient in lowering cost: The main personnel came from the civilian bureaucracy, the infrastructure was based on existing facilities like granaries and courier stations, and labourers were recruited according to household registers.

The frequent wars, often enough only fought to nourish the military, consumed annually more than two thirds of the state treasury. In the long run the need to cover running cost led to a lack in finances allocable for a modernization of weaponry or for the civilian infrastructure. After the campaigns of expansion which fixed the borders of the empire ended in 1795, warfare continued, but on a different level: The financially and morally disastrous war against the rebels of the pseudo-Buddhist White Lotus sect (*bailianjiao*) between 1796 and 1805 marked the beginning of a series of suppression of insurgencies throughout the empire (see Appendix 2). These revolts culminated in the foundation of the 'Heavenly Kingdom' by the Taiping rebels in 1851. Their parallel state endangered the existence of the Qing dynasty and could only be brought down by a thorough restructuring of the army, massive foreign aid in material and financial means, and a reform of the tax system.

75 During the sixteenth and seventeenth centuries the Ottoman government budget showed a surplus, and net balances remained positive in spite of continuous warfare. Murphey, *Ottoman Warfare*, p. 50. Yet this changed a century later, and a lack of financial resources led directly to the Ottoman defeat against Russia in 1783, compare Ágoston, *Guns for the Sultan*, p. 204.

This study is intended to contribute to the understanding of mechanisms, factors and instruments in logistics and war finance. It refers to cultural and institutional differences with other war-waging states, but also to common ground. The findings of this research are supposed to help explaining how the expansion of the Qing empire was possible and in what way various regions were integrated into the territory that is today called 'China'. The relevance of the bureaucratic apparatus in warfare is clarified, as well as the relation between the state and the private market, or organizations such as the salt merchant associations. It can be seen that the expansion of the Qing empire was a success story not only for operational, strategic and tactical reasons, or because the Manchus were a very belligerent people, but also because the government of the Qing—the emperors and their State Council—made use of a wide range of actors and interest groups to pursue the 'public cause' (*gongshi*). It kept a balance between merchants and state, regions and central government, as well as the military and the civilian population. This highly effective balance was codified, but in such an incomplete way that it did not engage the real problems in society (peacetime pay of soldiers), finances (an apparent abundance of state reserves), the administrative structure (professionalization in the civilian bureaucracy), and even war (naval warfare was not included). The 1780s can therefore be seen as a watershed between the perfect age of Qianlong, who fulfilled the historical mission of the Manchus as a conquering people, and his successors who were over-burdened with the attempt of homogenizing a society that had substantially changed since the days when the Manchus had first conquered China.

Sources and Methods

This book is divided into three chapters, covering the topics war expenditure, war finance, and logistics. The beginning of each chapter provides an introduction into the issue, followed by an analysis of the Qing implementation of the three particular themes from 1650 until 1800. The second half of each chapter consists of a discussion of the relevant topic during the second Jinchuan campaign.

The first parts of the three chapters are mainly based on the excellent and highly useful studies of Chen Feng and Lai Fushun.[76] Chen Feng

[76] Chen, *Qingdai junfei yanjiu*. Lai Fushun, *Qianlong zhongyao zhanzheng zhi junxu yanjiu* (Taibei: Gugong bowuyuan, 1984).

provides an in-depth study of the practice of military finance during the Qing period. He explains the military system of the Qing and the income of the soldiers in peacetime and during war. For the present book, his chapter on war expenditure was particularly helpful and provided an abundance of data otherwise only to be found in archival sources. Chen assembles a lot of empirical data to underline his arguments, but arranges these to fit thematical chapters, and not around individual campaigns. That type of work has been done by Lai Fushun, but exclusively for the Qianlong reign. He gives a short account of background of the so-called "Ten Successful Campaigns" (*shiquan wugong*) of the Qianlong reign and investigates for each of these wars the procedures of taking the troops to the war theatre and supplying them with grain and equipment, as well as the issues of mail service, and, somewhat briefly, finance and accounting.[77] His book is extremely helpful for the question of logistics, because he, too, makes use of a lot of not easily accessible archival sources.

My own study on financing and supplying the Jinchuan campaign is based on sources of three different types. Firstly, the official military annals (*fanglüe*) of the war,[78] the *Pingding Liang Jinchuan fanglüe*,[79] which includes selected imperial edicts processed for the general public, as well as the "Veritable Records" of the Qianlong emperor, *Gaozong shilu*,[80] which also mainly consist of edicts and can be used as a cross-reference for the *Fanglüe*; secondly, archival sources to which I had access during a visit in the First Historical Archives in Beijing (*Di yi lishi dang'an guan*), namely the "Draft copies of memorials of the State Council" *Junjichu lufu zouzhe*; then digitized memorials in the "Grand Archives of the Six Boards" *Neige daku dang'an* kept in the Academia Sinica (*Zhongyang yanjiu yuan*) in Taipeh, and the collection *Jinchuan dang* "Files from [the second] Jinchuan [war]", which includes a huge number of memorials, edicts and miscellaneous material (lists of officers to be rewarded, interviews with captives) from the archives; thirdly, I availed myself of a

77 The Ten Successful Wars were two campaigns in Dzungaria (1755, 1757), a suppression of the uprising of the two Khojas in Eastern Turkestan (1757), a series of campaigns against Myanmar (1765–1769), the two Jinchuan wars (1747–1749, 1771–1776), Taiwan (1788), Annam (1789), and two campaigns against the Nepalese Gurkhas that had invaded Tibet (1791, 1792).

78 On the genre of the 'military annals', see Ernst Haenisch, "Das Ts'ing-shih-kao und die sonstige chinesische Literatur zur Geschichte der letzten 300 Jahre", in *Asia Maior* VI (1930), p. 425.

79 *Pingding Liang Jinchuan fanglüe* (Siku quanshu edition).

80 *Da-Qing lichao shilu*, part *Gaozong Chun huangdi shilu* [henceforth abbreviated as *Gaozong shilu*] (Taibei: Huawen shuju, 1964).

collection of regulations, lists and official correspondence dealing with the issue of logistics and finance of the second Jinchuan war, namely the *Pingding Liang Jinchuan junxu li'an* "Archived precedents from the war against the Two Jinchuan",[81] compiled by Zheng Qishan (late Qing), who assembled and arranged a vast amount of relevant documents on the issues of logistics and finance from the Jinchuan war. To this source, the "War Supplies and Expenditures Code" *Junxu zeli* has to be added which was compiled after and based on the experience of the second Jinchuan campaign, as an empire-wide valid regulation for war expenditure. Its advantages and shortcomings will be discussed throughout this book.

Quantitative data have been extracted mainly from archival material and the prepared edicts that normally respond to palace memorials submitting information about particular problems, and often quote interesting specifications directly derived from practical experience. On the other hand, imperial edicts sometimes mention very crude and occasionally exaggerated figures that cannot be taken as written. Statecraft regulations as the various *zeli* 'precedent rules' of the ministries of Revenue (*hubu*) and Works (*gongbu*) also provide quantitative data about amounts of material and money. Such data, especially prices for commodities and services, have a jurisdictional character as ceiling amounts and therefore, too, cannot be taken at face value. Real market prices might deviate substantially from those fixed by the authorities, and constantly played an important part for the essential question of reducing war expenditure as much as possible.

81 Zheng Qishan, *Pingding Liang Jinchuan junxu li'an*, ed. by Xizang shehui kexue yuan, in *Xizangxue Hanwen wenxian huike*, vol. 2 (Beijing: Quanguo tushuguan wenxian suowei fuzhi zhongxin, 1991). This book will be abbreviated as *Jinchuan junxu li'an* also in the body of the text, in order to avoid confusion with the military annals *Pingding Liang Jinchuan fanglüe*.

CHAPTER ONE

WAR EXPENDITURE

> Recently accounts drop in from every division, claiming money for the production of tools without former notification, with the excuse that these were urgent matters. Others were not counter-checked and money was claimed back twice for the same thing. It also happened that officials did not reduce excessive sums, or mixed up bills and produced a lot of profit for themselves, to such a dimension that much money was squandered and embezzled in various ways.[1]

Each war of expansion of the early and high Qing period, either domestic or directed towards other territories, cost a tremendous amount of money. It is said that about seventy to eighty per cent of the annual revenue was spent for the military.[2] The total sum spent for a single war depended on the amount of troops brought to the war theatre, the duration of the campaign and the topographical circumstances, but also on how the campaign and the logistics were managed. The latter point is especially true for the campaigns in the late eighteenth century when large amounts of funds were misappropriated by commanders. To point out such cases directly is a matter of pure chance,[3] but wilful protraction of a campaign was known as a method to line the pockets especially of officers who were able to profit from rations and funds paid out only during a campaign and not under peacetime conditions.[4] The government even tolerated that the

1 *Pingding sanni fanglüe* (Siku quanshu edition), juan 35, fol. 9b–10a (KX 17/2/gengxu).

2 Chen, *Qingdai junfei yanjiu*, p. 9. Dai, "Yingyun shengxi", p. 6. Huang Tianhua, *Zhongguo caizheng shigang* (Shanghai: Shanghai caijing daxue chubanshe, 1999), p. 327.

3 See, e.g., a case of (intentional) mismanagement in *Gaozong shilu*, juan 321 (QL 13/r7/jisi), fol. 10a–11a. The example quoted above also shows that during the Kangxi reign the government was still not able to control the flow and use of funds during a war. Nancy E. Park, "Corruption in Eighteenth-Century China", in *Journal of Asian Studies* 56/4 (1997), p. 997, has pointed at a corruption case in the first Jinchuan war. It is known that officers also regularly confiscated funds belonging to subordinates or made money by passing on nags to common soldiers while pocketing the money they claimed for good horses. Dai Yingcong, "To Nourish a Strong Military: Kangxi's Preferential Treatment of His Military Officials", in *War and Society* 18/2 (2000), p. 89 (repr. in *Warfare in China Since 1600*, ed. by Kenneth Swope [London: Ashgate, 2005], pp. 51–72).

4 Dai, "Qing State, Merchants", pp. 55, 57.

military personnel were guaranteed adequate economic resources to enjoy a living standard 'on par' with the civilian officials who were entitled to other sources of personal income in the frame of their daily business.[5]

This chapter will explain what types of cost accrued during a military campaign, and to which of them the greatest part of the expenses were allotted. Albeit these types of cost did—generally seen—not differ very much from those of campaigns in other parts of the world, some items were characteristic for the Chinese approach of paying off troops and civilian labourers, for example, the 'baggage pay', or methods of reward. The person-related war-time payment to the military was much higher than the running cost for the army during peace, probably in order to forestall mutiny.[6] It is important to point out the high grade of bureaucratization that was involved in the procurement of objects and the accounting methods. Regulations for financial matters during each particular campaign ensured standardized accounting methods and a quicker auditing of the expenditure by the Ministry of Revenue (*hubu*). The great bulk of expenditure was, as will become evident, spent for transport along the logistics lines. This circumstance underlines the importance of the question of supply in warfare. It is also shown that the considerable expertise accumulated in the numerous campaigns ensured a quicker management of standard situations so that wars, waged by professional troops and organized by experienced officials of the civil administration, practically became part of the day-to-day routine of the dynasty's bureaucracy and therefore could be waged at lower cost. In spite of all attempts to regulate the cost of war by red tape there remained still excess cost not foreseeable by the war expenditure managers.

It is, especially for the earlier campaigns of the Qing period and even the Qianlong reign (1736–1795), not easy to provide exact figures on the expenditure for particular items, and in many cases the total sum paid for a campaign is not even known. At least for the most expensive of the Qianlong emperor's wars, the second Jinchuan campaign (1771–1776), a considerable amount of data is available to provide a better insight into the breakdown of cost. Because of the above said comparisons between individual campaigns which might resolve the question of a change in the cost structure, caused either by different conditions or by increasing mismanagement, are not possible, at least not in a systematic way.

[5] Dai, "Nourish Strong Military", p. 71.
[6] Dai, "Military Finance", p. 304.

In the first part of this chapter, general figures for the cost of early and high Qing period campaigns will be presented and briefly discussed. They are based on Chen Feng's excellent and detailed study on the war expenditure of the Qing (ending with the White Lotus rebellion).[7] In the second part, the total cost of the second Jinchuan campaign and its breakdown will be discussed in detail. That part will also describe some problems that occurred during the auditing process after the campaign.

War Expenditure in the Seventeenth and Eighteenth Centuries

Data for the first three reigns of the Qing dynasty (Shunzhi, Kangxi, Yongzheng) are fragmentary, but at least a few examples can demonstrate how substantial war expenditure could be.

For the Shunzhi reign (1644–1661) there are no clear figures of concrete campaigns, only such of running cost for the military in all regions. According to various sources, including data from the Ministry of Revenue, the annual military expenditure was about 13 million *liang* in the mid-Shunzhi reign and increased drastically from that time on. Some early experts like Wei Yuan (1794–1857) spoke of 24 million *liang*, but in some years, expenses of 30 million *liang* and more were necessary to meet the demands of the military machine. The problem with these figures is that both running cost for the military apparatus and cost for war were subsumed to one heading. Chen Feng estimates that the total expenditure for the military during the short Shunzhi reign might have reached 100 million *liang*.[8]

The Kangxi emperor had to wage several decisive wars that greatly contributed to the consolidation or even the survival of the dynasty, namely the elimination of the Three Feudatories (1673–1681), the campaigns to eliminate the Zheng 'dynasty' in Taiwan, and the final conquest of that island (1676–1683), the war against the Cossacks of Albazin (Yaksa, 1685–1686), the war against the Dzungars under Galdan (1696–1697) and the conquest of Dzungaria which was under the control of Tsewang Rabdan (1715–1721).[9] The conquest of Tibet (1712–1718) and Qinghai (1721–1723) as

[7] Chen, *Qingdai junfei yanjiu*.

[8] Chen, *Qingdai junfei yanjiu*, p. 242.

[9] The Dzungars (from Mong. *Jegün γar* 'left wing') were a sub-group of the Oirats (Oirad, Ölöd, Eleuths), the Western Mongols. The other three of the so-called four Oirat tribes were the Khoshots (Qošod, Qošuud), the Torgud (Torguud), and the Dörbed (Dörböd, Dorbet). Transcriptions of Mongol names and terms differ widely.

well as the suppression of numerous internal rebellions have to be added to this list.

The war against the Three Feudatories was the Kangxi emperor's attempt at getting rid of the all-powerful generals Wu Sangui, Shang Kexi and Geng Jingzhong, defectors of the Ming dynasty that had supported the Qing in their initial conquest of China. Enfeoffed as 'princes' (*wang*) in the southern regions of China these three generals established virtually independent kingdoms. The war against them lasted eight years and devastated almost all provinces of southern China. The Qing dynasty brought into the field more than 400,000 troops from all provinces, and almost 68,000 troops were newly recruited from among the population. Fortunately, a few data have survived that allow a rough estimation of the total cost of that war. Prince Le'rjin, highest commander of the army in Hubei, for instance, was able to produce an income of 3.95 million *liang* for the war chest, 1.87 million of which were derived from the field tax of various provinces. 300,000 *liang* were the income of the salt tax in Zhejiang and Liang-Huai, 400,000 *liang* came from the treasury of the Ministry of Revenue, and the rest were 'miscellaneous items' (*zaxiang*).[10] If the annual expenditure for war was about 4 million *liang* in the central war theatre of Hubei alone, the eight years' conflict cost about 30 million *liang*—only in that province, and in total probably three times that much. Surviving figures from the treasury of the Ministry of Revenue clearly demonstrate that during the Kangxi reign the annual amount of money in treasure that was not spent (*cun yin*) was about 20 million *liang*, with an increasing tendency, yet during the years of the war against the Three Feudatories, the remainder shrank to 5 or even 3 million *liang*. With a kind of extrapolation it can be estimated that some 50 million *liang* are missing, in other words, they were spent for the war.[11] The effective total cost of the war against the Three Feudatories was probably even as much as 100 or 150 million *liang*.

The war against the Dzungar leader Galdan had been well prepared so that sufficient amounts of grain were stored to be taken to the advancing armies along the logistics trails. The Kangxi emperor personally led an army of at least 80,000 troops. The cost of these campaigns is not easy to determine because sufficient data are lacking. Yet considering the length

---

[10] Quoted in Chen, *Qingdai junfei yanjiu*, p. 246. Chen calculates a sum of 3.5 million *liang*, probably leaving out the remissions from the Ministry's treasury.

[11] Quoted in Chen, *Qingdai junfei yanjiu*, p. 247.

of the campaigns and the cost of feeding the troops and their animals, paying for transport, providing weaponry and paying the troops their 'baggage pay' (*xingzhuang yin*), the first two campaigns might have cost about 3 million *liang* each, and the third campaign 10 million. The experience of these campaigns made estimation in advance possible, so that in 1715 the Kangxi emperor assessed the cost for the new campaign against Tsewang Rabdan as 3 to 4 million *liang*. Similar figures were still used for the western campaigns during the Qianlong reign for the previous assessment of cost. Not included in the war expenditure is the cost for occupation troops remaining in the war theatre. That campaign furthermore included the conquest of Qinghai and Tibet, whose aim it was to forestall the influence of the Dzungars on the Tibetan potentates in Lhasa. Chen Feng therefore estimates that the cost of this war in the late Kangxi years approached 50 million *liang*. It brought down the surplus in the ministerial treasury from 47 million in 1719 to 23 million *liang* at the beginning of the Yongzheng reign.[12]

The campaign against the fort of Albazin at the border to Russian Siberia was prepared long in advance, and although it required the construction of 144 boats, the cost did not surmount 1 million *liang*. Data about the conquest of Taiwan differ because the military activities against the Zheng 'dynasty' (Koxinga, or Zheng Chenggong, and his successors) began long before. The actual campaign, the conquest of the Penghu archipelago and the main island that occurred in 1683, was undertaken with a minimum of manpower and funds because the war against the Three Feudatories had virtually emptied the state treasury. A lot of weaponry, ships and rewards for the troops were therefore organized in the form of contributions (*juan*) from private persons to the common cause. High officials like the governor-general (*zongdu*) of Fujian, thus privately paid for the war. Chen Feng estimated the total cost of this campaign as 4 million *liang*. The uprising of Zhu Yigui in 1721 made a second conquest of the island necessary. The campaign itself was quite short, but in the aftermath it took a further two years to eliminate the last adherents of the rebel. The high grain price, the necessity to hire private ships, organize weaponry and recruit auxiliary troops led to expenditure to the amount of 9 million *liang*.[13]

During the short Yongzheng reign (1723–1735) more than 60 million *liang* were spent for campaigns. The most important wars were fought

---

12 Chen, *Qingdai junfei yanjiu*, p. 254.

13 Quoted in Chen, *Qingdai junfei yanjiu*, p. 245.

against the Dzungars and against the rebellious Miao tribes in the southwest. The campaign against the Dzungars in 1729 and 1730 devoured 3.9 million *liang*, that in 1731 and 1732 0.8 million *liang*, and expenditure for the campaigns in Uliyasutai (Outer Mongolia) between 1733 and 1735 accrued to 1.3 million *liang*. Detailed figures are mentioned in a memorial from the Qianlong reign stating that the western campaigns of the Yongzheng emperor consumed 54.4 million *liang*.[14] The war against the Miao tribes in Yunnan, Guizhou and Guangxi can be considered as being on the same level of cost. The two campaigns undertaken against the native population cost at least 3 million *liang*, with 4 million being more likely.

For the Qianlong reign, data are more abundant, but figures differ in various sources because even the official history of the Qing dynasty, the *Qingshigao*,[15] used to quote figures according to 'common knowledge' instead of relying on archival sources. Chen Feng has detected figures in a source called *Sichuan buzheng lu* "Records of the financial administration of Sichuan", which are more exact than those in the *Qingshigao*, the 'brush-notes' style book *Yanpu zaji* by Zhao Yi,[16] or Wei Yuan's eulogy on the dynasty's military, the *Shengwuji*.[17] Lai Fushun reports very exact figures based on archival sources and is even able to discern between sums that were allocated for the campaigns (*bokuan*) and those that were really made use of and were brought to account (*baoxiao*).[18]

The largest difference in these figures can be seen in an amount of 20 million *liang* that is mentioned in the *Qingshilu* as expenditure of the first Jinchuan campaign (1747–1748), and figures of between 7 and 10 million *liang* mentioned in other sources. This figure of 20 million is in some sources justified as including expenditure for disaster relief in Jiangsu, Zhejiang and Anhui.[19] The Qianlong emperor himself speaks of 20 million *liang* in an edict issued before the end of the campaign, when not yet any final account had been drawn up.[20] The figures used in the political rhetoric in his edicts cannot always be taken at face value, especially at the end of the first Jinchuan war, when he was highly enraged about the

---

[14] Quoted in Zhuang Jifa, *Qing Gaozong shiquan wugong yanjiu* (Taibei: Guoli gugong bowuguan, 1982; repr. by Beijing: Zhonghua shuju, 1987), p. 494. Chen, *Qingdai junfei yanjiu*, p. 258.

[15] *Qingshigao*, comp. by Ke Shaomin (Beijing: Zhonghua shuju, 1977).

[16] Zhao Yi, *Yanpu zaji* (Beijing: Zhonghua shuju, 1982).

[17] Wei Yuan. *Shengwuji* (Sibu beiyao edition).

[18] Lai, *Qianlong zhongyao zhanzheng*, Tables 49–50, pp. 426–430.

[19] Zhuang Jifa, *Qing Gaozong shiquan wugong*, p. 128.

[20] *Gaozong shilu*, juan 331, fol. 31b (QL 13/12/xinchou).

Table 1. War expenditure during the Qianlong reign according to different sources (in million *liang*).

	Qingshigao	*Yanpu zaji*	*Shengwuji*	*Sichuan buzheng lu*	Archival: allocated	Archival: used
First Jinchuan	20.000	7.750		7.605	9.964 [11.911]	7.127
Dzunghars, Khodjas	33.000	23.110	33.000		35.000	23.110
Myanmar	9.000	9.110	9.000		13.201	*9.118
Second Jinchuan	70.000	63.700	70.000	59.832	62.700	53.517
Taiwan		10.000	8.000		8.950	3.710
Annam					1.364	1.364
First Gurkha				1.000	1.500	0.600
Second Gurkha	8.000			11.000	7.750	2.856
White Lotus	200.000 [150.000]		100.000			

Source: Chen, *Qingdai junfei yanjiu*, pp. 261–262, 275.
* Source error in digits; [] corrections by Chen.

disastrous developments in the headquarters, and unfavourably compared the military and financial disasters of that campaign with his father's and grandfather's successful wars in the steppe. Similarly, the cost of the second Jinchuan campaign was in rhetorical arguments given as 70 million *liang* although it had practically cost 63 million 'only'.

The expenditure for the two campaigns in Dzungaria and that against the Muslim Khojas cannot be identified separately. Sources always name a summed-up figure of about 33 million *liang* for all three campaigns in the western regions. According to the military annals of the Dzungar campaigns, the first one of these wars cost 17 million *liang*, at least up to that point of time (1756).[21] In the same source the cost of the Yongzheng emperor's western campaigns are estimated as 50 to 60 million *liang*.

The figure of 13 million *liang* of funds allocated for the Myanmar campaigns includes money to be paid for post-war reconstruction (*shanhou*) as well as sums spent for constabulary actions before the actual beginning of the campaign and between the individual campaigns.

[21] *Pingding Zhungar fanglüe* (Siku quanshu edition), *Zhengbian*, juan 34, fol. 6a (QL 21/11/gengxu).

Concerning the second Jinchuan campaign, figures differ somewhat although relatively detailed information is available. This is to be attributed to numerous unsolved cases in the final account. Misappropriation and erroneous reports made a clear picture of the concrete expenditure impossible. Six million *liang* of expenditure of a total 63 million were just missing and could not be assigned to a concrete item.

Concerning the Taiwan campaign Chen Feng rectifies Lai Fushun's data about the funds used and argues that Lai's figures only include the expenditure of the war logistics bureau (*junxuju*) in Taiwan and not the cost that accrued in the bureau located on the mainland.[22]

The most serious problem concerning war expenditure of the early and high Qing period is the long-lasting campaign against the White Lotus uprising in the border areas of the provinces of Sichuan, Hubei, Hunan, Shaanxi and Gansu. In the beginning the war was waged as a constabulary action to put down local rebels and only gradually deteriorated into a 'war' that involved regular troops of several provinces. Chen Feng quotes several memorials and edicts that attest a figure of "more than a hundred million *liang*", sums up figures mentioned in edicts of the early Jiaqing reign and comes to the conclusion that the figure of 200 million *liang* listed in the *Qingshigao* is about 50 million too high.

Some general problematic issues in the accounting procedures concerning war expenditure have to be highlighted. The first one is the unclear treatment of issues included in the 'regular expenses' (*zhengxiao*) and those in the 'extraordinary expenses' (*waixiao*). Items included in the regular expenses file were paid for according to the War Expenditure Code (*Junxu zeli*). If expenditure surpassed the limits of these rules, e.g. expensive grain, it had to be included into the file of 'extraordinary expenses', even if the emperor had agreed to pay the higher price (*zhun xiao* 'allowed to be brought to account') because the items or service financed with this money were considered as indispensable for the campaign. The state then paid the regular price, while extraordinary expenses were to be covered by the responsible officials themselves, be it the highest commanders, the commander of the respective unit, or local officials. In many cases extraordinary expenses were laid on the shoulders of several responsible persons that commonly bore this sum (*fenpei*) and were so practically dealt with as contributions (*juan*) of these officials, yet without granting them a reward for their share, as was common for the institutionalized

[22] Chen, *Qingdai junfei yanjiu*, p. 265.

contributions. For this approved overspending a special account had to be drawn up, the *waixiao* 'accounts beyond (government budget)'. Both approved and unapproved overspending were recorded in the accounting books because it was necessary to trace back these sums to the officials that were held liable for repayment. Instead of an official participation in the campaign, these surplus cost could also be covered by contributions from merchants or the local gentry, a method called *guikuan* 'rendering a sum to where it belongs (i.e. the state)', or by levying a special surplus charge to be paid by the population, the so-called *bangtie yin* 'subsidies'.[23] In some cases the emperor graciously waived the repayment of these debts and the government settled the bill. The result of such a clear division between regular and extraordinary expenses was that a substantial part of the war expenditure was shouldered by private persons directly and was not included in the total sum of the cost. Voluntary or requested contributions by private persons, be it grain, horses, weapons or money, was also not counted as income of the war chest, except when the government launched large-scale contribution campaigns. Yet it was, in the end, part of the factual expenses for the war.

A second uncertainty is the treatment of cost that occurred in the initial stages of a campaign, before it was openly declared as a war, and in its aftermath. Defence measures, the creation of a logistics network, the installation of courier stations, the transport of grain and the preparation of weapons were not directly included in the war expenses. Only when a regular War Logistics and Expenditure Bureau (*junxuju*) was set up, cost was registered as war expenditure. Expenditure accumulated earlier was collected in so-called 'primary files' (*chu'an*) and attached to the main files (*zheng'an*). Even this was not done for measures necessary after the end of the campaign (*shanhou*). When the troops returned to their peacetime garrisons, other troops remained in the war zone to undertake reconstruction work, to open military agro-colonies (*tuntian*),[24] when people affected by martial activities (peasants being paid less for their grain, young men forcibly recruited to carry goods and foodstuff) had to be allowed a kind of compensation, weapons had to be repaired or widows and orphans of

---

23 Chen, *Qingdai junfei yanjiu*, pp. 184–185.

24 The military agro-colonies (*tuntian*), literally 'scion fields', were an invention of the Han dynasty (206 BCE–220 CE) for the control of the newly conquered Western Territories (modern Xinjiang). They were later also made use of inside proper China to facilitate land reclamation in barely inhabited regions or in border regions, where military garrisons were established and could not be supplied from elsewhere.

soldiers killed in action be granted indemnities. Only the cost for 'occupation troops' could later be brought to account and submitted to the central government. All other expenditure had to be shouldered by the provinces or the local government. The provinces also bore the cost of feeding the troops marching to the war theatre and back (if they did not come from within the province), the organisation and payment of mules and horses carrying their baggage as well as that of the courier service.

Finally, grain was to be brought to account in volume and not by price. The real price for grain on the market could differ widely, depending on the harvest and other conditions like distance from the production zones. Therefore regular disputes occurred about the price at which the war expenditure bureau was allowed to purchase. The officially permissible price for 1 *dan* (103.5 l) of grain was 1.2 *liang*, which was quite low under conditions of increased demand. Officials therefore regularly risked overpayment for rice, facing the starvation (or at least, mutiny) of their own troops and the logistics personnel. During the Yongzheng reign, for instance, the purchase of 1.1 million *dan* of grain from the market cost the organizers 21.8 million *liang*, but the Ministry of Revenue only agreed to pay 19.4 million.[25]

The Cost of the Second Jinchuan Campaign

Total Cost

The introductory survey of the *Pingding Liang Jinchuan junxu li'an* "Archived regulations for war expenditure of the second Jinchuan war" gives a general idea about the cost of the most expensive war of the Qianlong reign:[26] The first Jinchuan war had cost the government 7,604,800 *liang* of silver, 767,200 *dan* (79,400 m^3) of rice, and 20,260 *dan* (2.097 m^3)

---

[25] According to a clearance account in the archives, quoted by Chen, *Qingdai junfei yanjiu*, p. 257. Note: There are several cases that Chen made errors in transcribing Chinese figures (including the *wan* '10.000' unit) into Western-style ones, for instance, his Table 6–4, line 3, right column. There are also a few substantial clerical errors in his book, e.g. Wosheng'e 窝昇额 instead of the name Fengsheng'e 豐昇額. The cost of 21 *liang* per *dan* of rice in this example seems incredibly high if transport cost is not included. The original should therefore be checked for confirmation. Apart from the correctness of the decimals it is important to see that overspending was always immediately sanctioned. As a volume unit the character 石 is read *dan*, while the proper reading *shi* is only applicable to the weight unit, according to *Hanyu da cidian*, vol. 7, p. 980.

[26] *Jinchuan junxu li'an, Zonglüe*, fol. 1b–2b. Note that the title of this book will be, as already mentioned, henceforth abbreviated as *Jinchuan junxu li'an*.

of flour, as well as 7,440 *dan* (770 m^3) of beans for fodder. Everything was managed according to the experiences from the preceding campaign against the native king of Djandui (modern Garzê), in 1745. For the second Jinchuan war, the government had to allocate 61.6 million *liang* from the state treasury, 55.5 of which were really spent and brought to account. There were furthermore 3.8 million *liang* of tax abatements. The troops and labourers consumed 2,963,500 *dan* (306,700 m^3) of rice, the army used 4,271,400 *jin* (ca. 2,500 t) of gunpowder, and more than 3 million *jin* (ca. 1,800 t) of lead and iron for bullets and cannonballs. 650 iron cannon were cast, and more than 4,000 horses served in the courier stations in the war area. The overall cost, including those for further items like the construction of bridges, the building of boats and roads, as well as manufacturing every kind of tools, the imperial rewards for heroic troops and compensations for killed and injured soldiers, were "ten times as high as those of the former [first Jinchuan] campaign". Everything was done according to the preceding cases of the first Jinchuan war, the Myanmar campaigns and the campaigns in the Western Territories. It can be seen that as far as possible experience on organisation and finance from earlier campaigns was used. The final account of the second Jinchuan war comprised 47 chapters (*tiao*) with 870 issues (*an*). After a final check by the Ministry of Revenue the account was closed in 1785, almost ten years after the end of the war. It is mainly divided into the old files (*jiu'an*), covering the expenditure before August 1773, and the new files (*xin'an*), covering the expenditure after that date. The date was chosen because it was the watershed of the war, when the Qing army was utterly defeated in the uprising in the camp of Mugom in the summer of that year.

According to the general account in the *Jinchuan junxu li'an*, the Ministry of Revenue allocated more than 61.6 million *liang*, in fifteen instalments.[27] This sum is repeated somewhat more in detail at the end of the *Jinchuan junxu li'an*: The Ministry received or had at its disposal 61,731,754 *liang* that were allocated as war funds.[28]

The expenditure is itemized according to two criteria. The first one shows which of the three chronological types of file the expenditure belonged to; the second listing is according to the cost groups, or the purpose of the expenditure.

[27] *Jinchuan junxu li'an*, juan 2, fol. 177a.
[28] *Jinchuan junxu li'an*, juan 2, fol. 177b–178a.

Table 2. Expenditure for the second Jinchuan war (in *liang*) according to archival criteria and according to cost groups.

Expenditure according to archival criteria	
type of file	sum
'barbarian affairs' (*yiwu an*) and old files (*jiu'an*)	15,834,623.9145
new files (*xin'an*)	36,750,348.2135
additional expenditure	931,625.6683
first expenditure not included in other parts of the account	724.1150
'subsequent expenditure' (*xuxiao*)	2,703.3190
total	53,520,025.2403
Expenditure according to cost groups	
war-time pay for soldiers	7,543,010.8830
rewards and compensations	1,491,312.2690
staff in logistics stations, hiring craftsmen, and family allowances	368,830.5130
horses and labourers in courier stations	468,723.2050
transport of grain	38,633,984.3045
transport of equipment	4,018,719.2118
pay for civilian officials, constructing roads and bridges, and crafting boats etc.	912,172.5455
total	53,436,752.9300
difference to sum above	83,272.3000

Source: *Jinchuan junxu li'an*, juan 2, fol. 178a–b.

From this list it can be seen that about 30 per cent of the total expenditure was incurred during the first half of the war, while the second half of the war, from summer 1773, consumed almost 70 per cent of the total cost of more than 53 million *liang*. A very small amount of 724 *liang* spent at the beginning of the disturbances in Jinchuan was not entered into the regular account, but cleared separately. A further 2,700 *liang* were brought to account long after the war was ended and is called 'subsequent expenditure'. This sum was paid back by officials who were made liable for overspending.

It is really surprising that the highest cost did not accrue for paying the officers and the civilian officials dispatched to support the logistics. The 7.5 million *liang* for the 'baggage pay' and the provisions of the soldiers only constitute 14 per cent of the total cost, and even the 1.5 million *liang* for compensations and the rewards for them only amount to 3 per cent of the 53 million *liang* spent on the war. The staff in the logistics stations, the courier stations, and the workmen needed all over the war theatre, were paid so little that all this personnel did not require more than 0.84 million *liang*. The most expensive item was indeed the transport of grain to the logistics stations and the encampments, which required more than 38 million *liang*, or 72 per cent of the total cost. This sum includes the cost for the grain. The military equipment, of which only clothing and shoes are mentioned here, as well as the material needed for the cannons (i.e. iron, brass, gunpowder, etc.), was transported to the camps for an amount of no less than 4 million *liang*, or 8 per cent of the total cost. What exactly the 145 items (*an*) of equipment are, and if the price of the iron is included in the figures, cannot be found out because none of the items are detailed in the account. The synopsis of the expenditure for civilian officials under a chapter including cost for repair work and the necessary material, seems rather unusual. Thus it is a pity that it is not possible to find out more about the interesting question how much the civilian officials cost the government.

When comparing the sum of the two listings in Table 2 it can be seen that 83,272 *liang* are missing. This is probably the money spent at the beginning of the campaign, when the 'barbarian affairs' regulations were still in force, according to which ordinary constabulary activities against 'bandits' were arranged and accounted: The *Jinchuan junxu li'an* says that the items included in the first part of the table do not contain the expenditure from the 'barbarian affairs' (*yiwu*).[29]

The *Jinchuan junxu li'an* goes on listing the contents of the appended account (*fuxiao*). This account deals with sums which had been spent unduly and which were brought to account relatively late, when part of the money had been paid back as 'subsequent income' (*xushou*) by the responsible officials and merchants. When the final account was rendered, 1,543,814 *liang* were still missing, and the emperor renounced the payment of this sum.[30]

Adding up all those items, for which the emperor did not insist on the repayment of money due, the Ministry of Revenue came to the conclusion that the total cost of the war was 59,822,760 *liang*.

As could been seen above, more money was allocated than necessary (namely 61 million *liang*, only 53 of which had been spent), so that there

[29] *Jinchuan junxu li'an*, juan 2, fol. 178a.
[30] *Jinchuan junxu li'an*, juan 2, fol. 178b.

was a remainder (*shicun yin*) of 1,908,993 *liang*. Yet the greatest part of this sum was earmarked for issues concerning the border defence: 600,000 *liang* were to be used as pay for the garrisoned troops remaining in Jinchuan, 300,000 *liang* were stored in Dajianlu to protect the border garrisons in the south covering the road to Tibet, and 700,000 *liang* were budgeted for the provisional border defence. As for this project a total restructuring of the Jinchuan area was needed it was of the essence to keep ready the necessary amount of money. Thus only 308,900 *liang* were really left over.[31]

Yet there must have been some more remaining money. This can be seen in 1776, when a sum of 250,000 *liang* was assigned from Xi'an, Shaanxi, to the province of Gansu where natural disasters had caused a need for higher reconstruction funds. The money left over after the termination of the campaign (crudely said to be 3 million *liang*) was thus not totally sent back to the Ministry of Revenue or to other provinces, but remained in the large and important western garrison of Xi'an, probably in readiness for further military expenditure in case of need.[32] Part of the money stored in Xi'an might also have been repayments Sichuan had to hand over to Shaanxi, because the Xi'an garrison had provided weapons, gunpowder or other payments to the war chest in Chengdu, like the baggage pay for the many troops marching from Shaanxi to Sichuan. There is another document supporting the assumption that more money was left over from the Jinchuan war: In May 1776, 300,000 *liang* were sent to Guizhou from neighbouring provinces at least part of which came from the money left over from the Jinchuan war.[33]

From the final account included in the *Jinchuan junxu li'an* a lot can be learnt about the expenditure for miscellaneous items (*zakuan*).[34] Among them the accounts for rice, flour and beans are the most detailed and include the respective amount of foodstuffs and fodder transported along each one of the four logistics routes. Unfortunately such data are not available for all items, so that it is not possible to reconstruct exactly how much money was spent for all particular items.

In an edict, the emperor resumed the financial issues of the second Jinchuan war:[35]

31 *Jinchuan junxu li'an*, juan 2, fol. 180a.
32 *Gaozong shilu*, juan 1009, fol. 12a–12b (QL 41/5/26); juan 1011, fol. 1a–2b (QL 41/6/16).
33 *Gaozong shilu*, juan 1006, fol. 14a–14b (QL 41/4/8).
34 *Jinchuan junxu li'an*, juan 2, fol. 180a–187a.
35 *Jinchuan junxu li'an*, juan 2, fol. 187a–188b.

Table 3. Overview of the general account for the second Jinchuan war, income and expenditure (in *liang*).

entry	sum
money permanently kept ready for military expenditure (*jiuguan*)	1,100,000
money provided by the Ministry [of Revenue] and contributions (*xinshou*)	61,607,350
repayments by state officials (incorrect accounting, *xushou*)	613,754
total income	63,321,104
regular expenditure (*kaichu*)	53,240,476
money waived on imperial order	5,294,028
repayments to other provinces	4,485,925
total expenditure	63,020,431
balance (*shicun*)	300,672

Source: *Jinchuan junxu li'an*, juan 2, fol. 187a–188b.

The figures in this edict differ slightly from the statements found in the declarations before. The balance, for example, is given here as being 300,672 *liang*, while it was heard before that the balance amounted to 308,900 *liang*; the expenditure is said to have been 53,240,476 *liang*, while somewhat before we have heard of 53,436,752 *liang*, not including the expenditure incurred during the 'barbarian affairs' period at the beginning of the campaign. The accumulated value of the war chest was said to have been 62,741,554 *liang*, but here a sum of 63,321,104 *liang* can be found. One reason for the differences might be that grain, flour and fodder were counted separately and in the absolute volume units of *dan*, instead of converting them into the silver currency *liang*, a procedure which could lead to strongly diverging results, depending on the market price of the grain. The account reports a total amount of 3,074,796 *dan* (318,241 m^3) of grain, flour and beans shipped to the war theatre, and 3,058,909 of which were correctly brought to account. The balance was a mere 15,887 *dan* (1,644 m^3) of food and fodder.

As can be seen it is hard to reconstruct the particular entries in the final account of the war, because no single document is exact enough to show in what way the respective sums materialize. Yet even without knowing how they originated, it can be learnt how complicated it was to compile such a document as the final account for a military campaign that lasted more than four years, involved a dozen provinces and countless governmental and non-governmental actors. The entries in Table 3 also show how such an account was to be drawn up. There was a permanent budget available in the border provinces which served to support troops for a few months. This is called the 'old arrangements' (*jiuguan*). Because this sum of this budget was far from being sufficient to finance the whole war, the Ministry of Revenue and other governmental agencies, as well as private contributors, had to transfer considerable amounts of money to the provincial treasury of Sichuan, where it was budgeted for the war chest, as so-called 'new revenues' (*xinshou*). Each single item of expenditure (*kaichu*) had to be presented to the Ministry of Revenue, which checked if the expense was justified, or if there were instances of wastage or extravagance. The criteria for accounting were preceding cases from former wars, and especially the first Jinchuan campaign, during which the local circumstances were sometimes different from those in the Dzungaria wars (no horses could be used in the mountains; transport conditions were much more difficult; in winter, snow shovelers and adequate clothing were necessary).

Yet the experience of the second Jinchuan war and the problems arising when checking the accounts also contributed to the decisions which led to the compilation of the "War Supply and Expenditure Code" (*Junxu zeli*) that began immediately after the termination of the second Jinchuan campaign. Most accounts were justified, but there was quite a number of incorrect accounts, and in those cases the responsible officials had to refund the money out of their own pockets. In practice, it was deducted from their annual salary. Because the money wrongly brought to account had already been paid by the responsible officials, these sums must be counted as expenditure, but at the same time as income of the war chest, because they had to be refunded. There were a considerable number of financial items which the emperor graciously allowed to be written off, as could be seen in detail above. The last entry to be mentioned is the money that had to be sent back to other provinces having provided money to the province of Sichuan. In the first instance, this money was the 'baggage pay' for the troops coming from other provinces, but it also covered other equipment like clothes, gunpowder, bullets and arrows. The provinces were not expected to foot the bill for those items themselves, but were entitled to claim the money back from the war chest in Chengdu, respectively the Ministry of Revenue: In the résumé rendered in Table 3 the Ministry is mentioned as the sole large governmental contributor of the war finance, and not any of the provinces. Where the funds in fact originated will be seen in Chapter 2.

In the following sub-chapters the expenditure of particular items will be explained in detail. We will begin with the expenditure for troops, civilian officials, labourers and craftsmen, move on to the cost for food and fodder, horses and transport animals, the transport itself, and finish with various materials, from weaponry to construction materials. This chapter will be closed with a discussion of the problems of the auditing process.

Troops

The most important group of persons involved into the war was, of course, the military staff, that is, common soldiers and officers, who were, according to the traditional duality defined as 'military officials' (*wuguan*) in contrast to the 'civilian officials' (*wenguan*). Among the highest ranks commanding the troops there were also some non-combatants, who were obliged to go to the front because of their position within the administrative structure. Such people are, for example, the governor-general (*zongdu*) of Sichuan, but also officials from the central government who were delegated by the emperor to take over military duties, as, for instance, Grand Ministers Consultant (*jinglüe dachen*), or members of the State Council (*junjichu*) sent out to supervise the progress of the war as representatives of the emperor.

Besides these high-ranking civilians that, for the time of war transformed into soldiers, there were a lot of civilian officials from all over the country dispatched to the war theatre in order to take care of the organisation of supplies of all kinds. As opposed to the military staff, whose exact numbers are known, we are unaware of the exact number of civilian officials acting in Jinchuan.

By far the largest group involved in the war were the hired porters and labourers. Again, there are only approximate figures of how many people were hired to transport rice, gunpowder, bullets and all the other items necessary to wage a war. Yet at least some figures for the porters are known, while we are not able to estimate the number of craftsmen involved, as carpenters, ferrymen, cannon casters, as well as that of professionals like physicians or map drawers.

The soldiers taking part in the large campaigns of the Qing emperors came from three different types of military units. The largest proportion were regular soldiers from garrisons in various provinces, some from provincial capital garrisons, and others from smaller types of garrison throughout the provinces. This system of purely Chinese professional soldiers was called that of the 'Green Standards' (*luying* or *lüying*),[36] and was inherited from the Ming dynasty military system with its local garrisons (*zhenshu*). The total number of Green Standard soldiers throughout the country was about 650,000.[37] During the second Jinchuan war, Green Standard troops were deployed from garrisons in the provinces of Sichuan, Yunnan, Guizhou, Shaanxi, Gansu, Hunan and Hubei. The Green Standard troops are commonly called 'Chinese troops' (*Hanjun*, or *Hanbing*), sometimes also *luqi* or *lüqi* 'the Green Banners'.

Apart from this purely Chinese type of military troops, the rulers of the Qing dynasty installed a different, genuinely Manchu system of professional troops, the so-called Eight Banners (*baqi*). While the standards of the Chinese troops were green, as the name suggests, the standards of the Eight Banners were yellow, red, white, and blue, in two different designs, namely plain and bordered. Although a great part of the Banner troops were Manchus, there were also eight Chinese Banners, and eight Mongol Banners (adding up to a total of twenty-four Banners, of which the Chinese and Mongol Banners were of much smaller dimensions), but because each person could be shifted from one Banner to another, the ethnic composition of the Banners was often mixed. The name 'Manchu troops' (*Manbing*) is therefore misleading, because there were also many Chinese and Mongols serving in the Banner troops, and thus it seems better to use the term 'Banner troops' instead. While the garrisons of the Green Standard troops were scattered throughout the country, the Banner garrisons were concentrated at certain points of strategic importance. These were, of course, the empire's capital Beijing, where many guard units served to protect the imperial city and the imperial palace. In Beijing, some elite units equipped with modern weapons (according to seventeenth century standard), namely muskets and cannon, were garrisoned. Of great importance was further the homeland of the Manchus, the three northeastern provinces (or Three Eastern Provinces, *dong san sheng*) of Shengjing (later Liaoning), Jilin and Heilongjiang, and some strategic points in the newly conquered far west (the 'new territories', *xinjiang*), like Ili, or Xining (later Qinghai). In the other provinces, the Banner troops were quite few and

[36] Most of the important Chinese dictionaries, as well as Hucker, *Official Titles*, No. 3862, and Brunnert and Hagelstrom, *Present-day Political Organisation*, p. 337 (No. 749), prefer the reading *luying*. Yet there are also reasonable arguments for the reading *lüying*, see Xin Xingbao, "*Lüying* de duyin", in *Zhongxue lishi jiaoxue yanjiu*, 1/2 (2007), p. 88.

[37] For more information about the Green Standard troops, see Luo Ergang, *Luyingbing zhi* (Beijing: Zhonghua shuju, 1982).

concentrated in the provincial capitals or other important places. The Banner garrison in Chengdu, for example, was only 1,638 men strong, 32 of whom were workmen needed for the production of bows and arrows (*gongjiang, jianjiang*) or iron tools (*tiejiang*).[38] The Green Standard troops in Chengdu, on the other hand, consisted of 6,272 men.[39]

The third type of troops participating in the second Jinchuan campaign were native auxiliary troops (*tubing*) provided by native kings being loyal to the Qing government. Whenever the rulers of Jinchuan harassed their neighbours, the governor-general of Sichuan had ordered the kings of the region to dispatch a contingent of soldiers to calm down the insubordinate Jinchuan rulers. This kind of troops were not regulars within a standing army, but fighting units consisting of recruits called to the arms to fulfil their obligations towards their village heads and kings. Sometimes the native auxiliary troops were equipped with weaponry provided by the government of Sichuan, and the structure of their units imitated that of the Green Standard units. The War Expenditure Regulations *Junxu zeli* also discern non-Banner Mongolian troops as a distinctly paid group of fighters.

The task to determine the exact number of military personnel who had taken part in the campaign is quite easy. It is much more complicated to find out how many soldiers were fighting at the front at what point of time, or when contingents exactly left their home garrison and arrived in the camps. This must have been important for the calculation of the cost of paying and feeding the troops, but is a tremendously complicated matter because the exact numbers of troops constantly changed. Soldiers were killed in battle and could often not immediately be replaced by fresh troops from the home garrison. The standard size of military units (*ding'e*) was to be reached again by replacing the wounded and killed by new recruits (*xinmu bing*) or by the supplementary troops (*yuding, yubing*), which otherwise served as manservants (*genyi*, 'orderlies').[40] Wounded

[38] Added up according to figures in *Baqi tongzhi* (Changchun: Jilin wenshi chubanshe, 2002), juan 35, *Bingzhi* 4, p. 625.

[39] Added up according to figures in *Huangchao wenxian tongkao* (Hangzhou: Zhejiang guji chubanshe, 2000), juan 187, *Bingkao* 9, p. 6490.

[40] The word *yi* in *genyi* is an antiquated term originally meaning unpaid *corvée* labour, making it seemingly a kind of slave labour. In the Qing period all kinds of people recruited for service to the government were called *yi* (like *jiangyi* 'recruited craftsman'). Manservants were often younger brothers of soldiers that had not inherited the post of their elder brother, or sons of serving soldiers. As 'supernumerary troops' (*yuding*) they had the opportunity to become soldiers themselves which was not the case with manservants that were not members of military households. In some provinces the *genyi* were generally called *yubing* 'superfluous troops', or 'reserve troops', and *genyi* could be recruited from

soldiers were allowed to pause for some months to regain their full strength. In some cases, wounded high-ranking officers could be taken back to their home-garrison and civilian officials to their home-towns.[41] It is not clear whether supplementary troops, at least those of certain provinces, were paid as full soldiers or as manservants.

In total, troops from fourteen garrisons of different provinces were dispatched, as well as auxiliary troops provided by eighteen native kings. The *Jinchuan junxu li'an* confronts the modern scholar with two different figures, firstly 129,500 men of all three types of troops (*Man-Han-tu bing*), of which 14,731 were killed (i.e. 11.4 per cent),[42] and then, a few pages further on, with the figure of 145,126 men.[43] Of these, it is said, 47,795 men returned to their garrisons (in other provinces than Sichuan) at the end of the campaign, and more than 40,000 were wounded, or ill or had deserted. Over 27,000 men out of the total number were native auxiliaries. The difference of 15,600 troops in the two 'total sums' might be the result of an inclusion of a certain number of supplementary troops (*yuding*) into the total figure. This demonstrates the flexible character of the supplementary troops that could be used either as manservants, or as fighting units.

The intermediate sum of 114,133 in Table 4 is 15,367 men lower than the 129,500 men mentioned in the introduction of the *Pingding Ling Jinchuan junxu li'an*. Yet when adding the figure of 29,597 native auxiliary troops to the 114,133 Banner and Green Standard troops, a total figure of 143,730 men is reached, which is only 1,396 men less than the documented total number of 145,126 troops. This is the total number of participating soldiers of all types of troops, including the native auxiliary troops and the 10,200 supplementary troops (*yuding*) from Shaan-Gan and Liang-Hu.

It can be seen that 70 per cent of the troops were Green Standard troops, and that their units took over the main part of the fighting, at least in pure numbers. Yet the combat strength of the Banner units, which account for only 8 per cent of all troops, cannot be neglected. While many

among the number of troops surmounting the regulated standard-size of the garrisons. A common relation between regular troops and manservants was 30 supplementary soldiers for 100 troops. Yet there were also precedents from the western campaigns that 4 cavalrymen were given one supplementary soldier and five infantrymen one supplementary soldier respectively. It was in that case decided that troops who could not put up the respective number of supplementary soldiers should hire civilian manservants to carry their luggage. *Pingding Liang Jinchuan fanglüe*, juan 36, fol. 2b–3b (QL 37/7/jiyou). The Manchu term for *genyi* is *dahaltu*, which is a literal translation of the Chinese word, meaning 'following [servant]'.

[41] Compare *Hubu junxu zeli*, juan 6, fol. 6b–8a.

[42] Figures according to *Jinchuan junxu li'an, Zonglüe*, fol. 2a–3b.

[43] *Jinchuan junxu li'an*, juan 1, fol. 7a.

Table 4. Types, origin and proportions of the imperial troops during the second Jinchuan war.

origin		number
Banner troops		
Capital troops from Yunnan		263
Chengdu garrison		1,762
Capital and northeastern provinces		6,521
Jingzhou garrison		2,202
Xi'an garrison		2,041
	sum Banner troops	12,789
Green Standard troops		
Sichuan		25,267
Guizhou		15,388
Shaan-Gan		36,643
Liang-Hu		15,949
Yunnan		8,097
	sum Green Standard troops	101,344
	sum Banner and Green Standard troops	114,133
Native auxiliary troops		29,597
	total sum	143,730

Source: *Jinchuan junxu li'an*, juan 1, fol. 7a–15b.

Green Standard troops were quite poorly equipped, a musket was—besides the traditional bow and arrow—standard equipment of the elite Banner troops.[44] The native auxiliary troops, amounting 22 per cent of all troops, can almost be neglected in respect to their effectiveness. The official documents permanently speak of their uselessness and cowardliness, and it seems to be that the native auxiliaries only served as a kind of 'cannon fodder' in the first front.

Baggage Pay

The Chinese military system of the early and high Qing period was that of a group of professionals which were paid their salary (*fengxiang*) in peacetime and during war. When talking about the war time expenditure,

44 Dai, "Military Finance", p. 306.

this salary can therefore not be regarded as money spent for the military campaign.[45] Yet the regular salary of the military played a role for the calculation of the so-called baggage pay (*xingzhuang yin*) of the officials. This allowance served to support the soldiers when preparing their luggage and to care for all things necessary when leaving the garrison and going out to a campaign, to buy or repair uniforms, boots and weapons, which were normally not provided by the garrisons but were a private matter of each soldier.[46] This baggage pay was a stipend given according to the regular annual salary of officers, and was a one-time pay, which should enable the soldiers to make ready for deployment. Of course, a part of the money was also to be used to nourish the family which was left behind. 'Corporals' (*waiwei, lingcui*) and common soldiers were given a fixed sum of money, while the baggage pay for officers was geared to their salary as state officials.

The baggage pay was first meant to be a surplus stipend by the emperor, which had not to be given back, but from the Kangxi reign on the *xingzhuang* stipend was transformed into a loan for the soldiers, although it was often the case that after the campaign, the emperor dispensed with the back payment. In some exceptional cases, when a campaign was lingering on for a long time, it had been the case that the *xingzhuang* stipend was paid out for a second time.[47]

Until the final edition of the *Junzu zeli* was compiled in 1785, the practice during the various campaigns was very different, and the amounts of the stipends could exhibit large disparities. High officers could get a pay varying from one to two years' salaries, or only a stipend 100 of *liang*, and it was not prescribed if the money was donated or only given on loan. Unfortunately the commentary to the *Junxu zeli* does not explicitly say which regulations were in force for the second Jinchuan campaign ('the Sichuan precedents', *Sichuan li*),[48] yet they must have played a certain role for the suggestions made by the compilers of the *Junxu zeli*.

The baggage pay during the second Jinchuan war differed considerably for troops from different provinces. Different amounts of baggage

[45] For peace-time income of military personnel, see Dai, "Military Finance", pp. 298–302, as well as Dai, "Yingyun shengxi". The income of the Qing soldiery proved inadequate after the territorial expansion of the empire and the large conquest wars were ended in 1795.

[46] Dai, "Military Finance", p. 302.

[47] *Hubu junxu zeli*, juan 1, fol. 4a.

[48] *Hubu junxu zeli*, juan 1. The data are also included in Lai, *Qianglong zhongyao zhanzheng*, pp. 395–399, and Chen, *Qingdai junfei yanjiu*, pp. 50–51, 118–119.

pay were also granted to the various types of civilian and military staff (Banner troops from the Capital, civilian officials from the central government and the local governments, Banner troops from the northeast, from other provinces, Mongol non-Banner troops, Green Standard troops, and native auxiliary troops). Furthermore, the money given to Banner troops was intended to be a gratification, while the baggage pay for Green Standard troops should partly be granted for free, and partly be given on loan. Furthermore, the Green Standard troops were not given any baggage pay for their manservants.

Officers of the Green Standard troops received a two years' salary for baggage pay, while 'corporals' and common soldiers were given a fixed pay.[49] The salary upon which the baggage pay was based, was only the nominal salary (*fengxiang*) of the officers and not their actual annual income, which was much higher, when adding the so-called firewood allowance (*xinyin*), vegetable-candle-and-coal allowance (*shucai zhutan yin*), and the sealpaste-and-paper allowance (*xinhong zhizhang yin*).[50] From 1753 on all pay for military officials was given according to rank grading. Except those items which were nominally intended to be used for daily life and to manage the office, military officials, like civilian officials, were given a so-called anti-corruption allowance (*yanglian yin* 'pay nourishing incorruptibility'), which was intended to prevent corruption of officials caused by inadequate income. The peace 'price' of Banner troops for example can be found out through a calculation by an unknown official who estimated that 500 Banner troops cost 54,000 *liang* annually, according to other figures, 1,000 men cost 84,800 *liang* annually, which means, between 85 and 108 *liang* per person.[51]

After the first Jinchuan war the loans in shape of the baggage pay had been given as gratifications and were not to be paid back in annual instalments, as the *Junxu zeli* later prescribed.[52] To complicate matters further, only the troops of Shaanxi and Gansu were originally given an additional loan on top of their baggage gratification. After some discussion it was therefore, contrary to the precedents from the first Jinchuan war, allowed that the troops of all provinces could be given a loan to be subtracted

---

[49] Annual pay in accordance with Luo, *Luyingbing zhi*, pp. 342, 359, and Huang Huixian and Chen Feng, *Zhongguo fenglu zhidu shi* (Wuhan: Wuhan daxue chubanshe: 2005), pp. 632–633.

[50] See Huang and Chen, *Zhongguo fenglu zhidu*, p. 611.

[51] *Jinchuan dang* 37/IV/00107 (QL 37/10/16).

[52] Documented by Chen, *Qingdai junfei yanjiu*, p. 228.

Table 5. Comparison of baggage pay for Green Standard troops from different provinces during the second Jinchuan war (in *liang*).

	Sichuan	Shaan-Gan		Hu-Guang	Yun-Gui
rank	loan	gratification plus…	loan	loan	gratification
provincial military commanders (*tidu*)	1,000	2 years salary	350		
regional commanders (*zongbing*)	1,000	"	250	500	
regional vice commanders (*fujiang*)	500	"	200	300	60
assistant regional commanders (*canjiang*)	150	"	150	250	60
brigade commanders (*youji*)	150	"	150	200	40
brigade vice commanders (*dusi*)	100	"	100	100	40
assistant brigade commanders (*shoubei*)	100	"	100	80	30
company commanders (*qianzong*) and quad leaders (*bazong*)	40	"	40	50	15
corporals (*waiwei*)	12	"	12	40	8
common troops (*bingding*)	4–6	6–10	4	15–24	1.3

Source: *Jinchuan junxu li'an*, juan 2, fol. 37a–40a.

from their salary after the war was over.[53] Although the emperor had given his blessing to the plan of handing out loans to all Green Standard troops, no generally applicable numbers and amounts had been fixed at that time.[54] Table 5 shows how different the regulations for the different provinces were during the second Jinchuan war.

---

[53] *Jinchuan junxu li'an*, juan 2, fol. 40a.

[54] The regulations in the *Jinchuan junxu li'an* are somewhat confusing. In juan 1, fol. 38b, it is written that all officers of either type of troops (*Man-Han guanbing*) were given a gratification of twice their yearly salaries. Yet the same paragraph deals with gratifications only for Banner troops.

Concerning the native auxiliaries, the *Jinchuan junxu li'an* says that each person was given 3 *liang* of baggage pay and 2 *liang* of family allowance (*anjia yin*), which comes to a lump sum of 5 *liang*.[55] In the *Junxu zeli* this practice is reflected in the commentary which criticizes that it does not make any sense not to distinguish between ranks and officers (the latter actually village heads, chieftains, and kinglets). In the Myanmar campaigns the baggage pay for native troops was 6 *liang*, an amount which served as the basis for the new regulations for military expenditure in the *Junxu zeli*. Nevertheless the *Junxu zeli* commentary omits the existence of a family allowance, paid to the native troops deployed against Jinchuan, and which was normally only given to civilian labourers. The baggage pay and family allowances for native auxiliary troops were obviously not to be paid back. At least there is no such regulation in the *Junxu zeli*, and it would also have been quite difficult to enforce back payment because the native auxiliaries were not professionals garrisoned in a particular place which could easily be controlled.[56]

The most complicated question concerning the baggage pay is that of the Banner troops. Here regulations minutely distinguished between Banner troops from the Capital, troops from the northeastern provinces, and troops from the provincial garrisons. Apart from the soldiers, their manservants were also given a kind of pay for deployment, which was called 'leather clothing pay' (*piyi yin*).[57] On the other side, Huang Huixian and Chen Feng conclude that the salary of the provincial troops was not very different from that of the Capital troops.[58]

At the beginning of the campaign, the provincial Banner troops from Sichuan took with them a certain amount of the baggage pay (in that case called *jiaxiang yin*),[59] namely 40 *liang* for a vice commander-in-chief (*fudutong*), 4 *liang* for officers of the rank of assistant commandant (*xieling*), company commander (*zuoling*) and platoon commander (*fangyu*), 2.5 *liang* for lieutenant (*xiaojixiao*) officers and military clerks (*bithesi*), 2 *liang* for 'stout' (*zhuangda*) 'corporals' (*qianfeng*, *lingcui*), cavalrymen and gunners, and 1 *liang* for military workmen (*jiangyi*).

---

55 *Jinchuan junxu li'an*, juan 1, fol. 3b.

56 *Jinchuan dang* 39/III/00217 (QL 39/8/20).

57 *Hubu junxu zeli*, juan 1, fol. 2a–3a.

58 Huang and Chen, *Zhongguo fenglu zhidu*, pp. 605–606.

59 *Jinchuan junxu li'an*, juan 2, fol. 26a. The term *jiaxiang yin* might be a clerical error for *yuexiang yin* 'monthly pay', as the character *jia* can only be seen in the combination *majia xiangyin* 'pay for armoured cavalrymen'.

Compared to the Green Standard troops, of whom each individual of the common troops generated deployment cost of between 6 *liang* (infantrymen and guardsmen) and 10 *liang* (cavalrymen), the Banner troops were more expensive: Each provincial Banner man generated deployment cost of between 10 *liang* (cannoneers, heavy infantrymen, and professional craftsmen) and 20 *liang* (light and heavy cavalrymen and corporals), Capital Banner troops even between 20 *liang* (infantrymen) and 30 *liang* (cavalrymen).[60] The complaints that Banner troops were several times as expensive as the Green Standard troops can be well understood.[61] The baggage pay for both types of Banner troops (Capital and northeastern) differed, as can be seen, for example for the common soldiers (40 *liang* for Capital troops and only 30 *liang* for northeastern troops), and for the regimental commander (*canling*, 260 *liang* for Capital troops and only 180 *liang* for Shengjing, Heilongjiang and Jilin troops). In the *Junxu zeli* the baggage pay for the Capital and northeastern troops was called 'packing pay' (*zhengzhuang yin* 'pay for fixing the equipment'). The Capital officers were granted a pay of a two years' salary, the troops from the northeast were given a fixed amount of money. The Dzungar troops from Ili are included in these figures, although there existed some special regulations for the baggage pay and the number of manservants for non-Banner Mongol troops.[62]

An item not prescribed by the *Junxu zeli* are loans for the Banner troops to prepare their baggage. Although the regulations issued in the *Junxu zeli* are based on older precedents from several former wars, and only talk of gratifications, the case of the second Jinchuan war shows that it was not uncommon for the baggage pay for Banner troops to be given on loan and not as a gratification. The *Jinchuan junxu li'an* only talks of loans given to the Banner troops[63] and does not list any memorials which ask for a remission of the loans. During the western campaigns, the conditions had been much better: Each officer—Banner and Green Standard—had obtained a gratification of two years' salaries, and Capital Banner troops

---

[60] *Hubu junxu zeli*, juan 1, fol. 1a–4a. Lai, *Qianlong zhongyao zhanzheng*, pp. 118–119, provides concrete figures to support those prescribed in the War Expenditure Code *Junxu zeli*.

[61] *Pingding Liang Jinchuan fanglüe*, juan 20, fol. 14b–15a (QL 37/2/dinghai).

[62] *Hubu junxu zeli*, juan 1, fol. 3a; juan 3, fol. 2b–3a. Those regulations probably only concern 'proper' Mongol banners (the new leagues and banners, not the old Mongol Banners) in Outer Mongolia and not contingents of other Mongols which were part of other units, like the Oirat Mongols, who served under the General of Ili.

[63] *Jinchuan junxu li'an*, juan 1, fol. 37b–38b.

were given a baggage pay of 40 *liang*, provincial troops 30 *liang*, and Green Standard troops 20 *liang*.[64] These statements correspond to the later regulations of the *Junxu zeli*.

Salted-Vegetable Pay and Daily Rice Rations

Except the baggage pay, which was a lump sum paid out at the beginning of each campaign (except in some cases, when only half of the money was paid in the home garrison, and the other half later, upon arrival in Chengdu),[65] the soldiers were paid a certain sum of money per month or day to buy their food. The name for this allowance is therefore 'salted-vegetable pay' or 'salt-and-vegetable pay' (*yancai yin*), together with the money for daily rations (*kouliang yin*) often subsumed under the name of *yancai kouliang*. For all different kinds of professionals taking part in the war, there were diverging regulations and practices for the daily rations. These could either be given in rice (*bense* 'original commodity'), or as a sum of money to buy food with (*zhese* 'converted commodity'), or partly given as rice and the rest paid out in silver. For the conversion of rice into money, the government had fixed an official conversion rate, which is at 1.2 *fen* (0.012 *liang*) of silver for 1 *sheng* (ca. 1 litre) of rice, or 1.2 *liang* for 1 *dan* (100 *sheng*) of rice, or the other way round, 0.85 *sheng* of rice for 1 *fen* (0.01 *liang*) of silver.

The salted-vegetable pay, which was given to buy other food than rice, was 0.01 *liang* per day for each soldier, but nothing for the manservants, thus the soldiers and officers had to care for their manservants and to buy additional food for them from their own salted-vegetable pay. To make things more complicated, on the southern route (from Dartsedo northwards), the soldiers were given 0.015 *liang* per day from QL 36/8/15 (Sep 23, 1771) on.[66]

For the native auxiliary troops, the same regulations applied in the beginning: 0.0135 *liang* of silver per day to buy rice when inside the country, and 1 *sheng* of rice or 1 *jin* (597 g) of flour when outside the country.[67] The reason for this is that it is easier to supply rice than to arrange a possibility to buy rice with money—which then both (rice and cash) have to

---

64 *Jinchuan junxu li'an*, juan 1, fol. 38b.

65 *Pingding Liang Jinchuan fanglüe*, juan 35, fol. 13b–14a (QL 37/7/jiachen).

66 *Jinchuan junxu li'an*, juan 1, fol. 3a–3b.

67 In Europe, soldiers of the same age were entitled to two pounds of bread per day, and one pound of other food, see van Creveld, *Supplying War*, 21, 34. Lynn, *Feeding Mars*, p. 20, estimates the need of flour as 10 kg per day.

be provided. Under the conditions of increased demand the market price of rice was, furthermore, substantially above the official conversion ratio. Soldiers therefore preferred directly be given rice instead of money to purchase over-priced grain.[68] Upon arrival in the camp, the troops were daily given a salted-vegetable pay of 1 *fen* (0.01 *liang*), on the southern route 1.5 *fen* (a supplement of 0.5 *fen*/day to rent a porter). Additionally, native auxiliary troops received a monthly encampment pay (*zuoxiang yin*) of 9 *qian* and monthly 3 *dou* (31.05 l) of rice. Although the commentary to the *Jinchuan junxu li'an* says that part of this money was not paid for a certain time and therefore problems in the accounting had arisen, there is documentary evidence that the encampment pay was paid out throughout the whole campaign,[69] a fact which is also reflected in the commentary to the *Junxu zeli* which says that native auxiliary soldiers and 'corporals'—not the officers—obtained a monthly encampment pay of 1.155 *liang* of silver (0.9 *liang* plus some 25.5 *fen* [0.255 *liang*] to buy rice). The native chieftains and village heads, who acted as native officers, were not eligible for this payment.[70]

Later on, when the campaign developed into a war and the 'barbarian affairs' were transformed into military affairs, the regulations for the monthly and daily provisions became more complicated because there were no generally applicable rules for the whole empire, so that each province followed a different practice how to nourish their soldiers during war.[71] Only when the *Junxu zeli* was issued, the rules described in that canon became applicable throughout the empire.

The *Jinchuan junxu li'an* explains all the details concerning the salted-vegetable pay for troops coming from different provinces, and in the commentary to the *Junxu zeli* the respective amounts of pay for troops in the Sichuan (Jinchuan) regulations are cited.[72] A comparison between the accounting customs at the time of the Jinchuan campaign with those later validated empire-wide in the *Junxu zeli* can demonstrate how cost was cut later on. The number of manservants each officer and soldier was allowed was also different in some cases.

---

[68] Dai, "Military Finance", p. 302.

[69] *Pingding Liang Jinchuan fanglüe*, juan 57, fol. 17b (QL 38/r3/renwu). See also *Jinchuan junxu li'an*, juan 2, fol. 32b.

[70] *Hubu junxu zeli*, juan 1, fol. 4b–5a.

[71] See *Jinchuan junxu li'an*, juan 2, fol. 25a–43b.

[72] *Hubu junxu zeli*, juan 3, fol. 3a–4b. A part of these regulations followed older precedents from the western campaigns.

Table 6. Comparison of salted-vegetable pay (in *liang* per month) and number of manservants of Green Standard troops and local civilian officials according to different regulations.

rank	pay according to Sichuan precedents	pay according to *Junxu zeli*	manservants according to Sichuan precedents	manservants according to *Junxu zeli*
Green Standard troops				
provincial military commanders (*tidu*)	15	12	32	24
regional commanders (*zongbing*)	15	9	24	16
regional vice commanders (*fujiang*)	7.2	7.2	16	12
assistant regional commanders (*canjiang*)	4.2	4.2	10	10
brigade commanders (*youji*)	4.2	4.2	8	8
brigade vice commanders (*dusi*)	4.2	3	6–8	6
assistant brigade commanders (*shoubei*)	2.4	2.4	6	6
company commanders (*qianzong*)	2.4	2	3	3
squad leaders (*bazong*)	1.2–1.5	1.5	3	3
corporals (*waiwei*)	0.9–1.3	1.5	1–2	2
common troops (*bingding*)	0.9–1.3	0.9–1.3	0.3	0.3
manservants (*genyi*) of officers	0	0	–	–
manservants of common troops	0.5	0.5	–	–
local civilian officials				
circuit intendants (*daoyuan, daofu*)	30	4.2	8	8
prefects (*zhifu*)	25	4 (?)	7	6 (?)
sub-prefectural (*tongzhi, tongpan*), department (*zhizhou*) and district magistrates (*zhixian*)	20	4	5	6
assistants (*zuoza*)	15	2.5	3	3
manservants (*genyi*)		0.5	–	–

Source: *Jinchuan junxu li'an*, juan 2, fol. 27b–28a, 41b; *Hubu junxu zeli*, juan 2, fol. 6a–7a; 3, fol. 3a–4b.

The regulations for the troops from the provinces of Guizhou and Yunnan were the following, according to the *Jinchuan junxu li'an*: From the home garrison to the camps, all officers from the rank of assistant brigade commander (*shoubei*) upward were not given any pay or provisions; company commander (*qianzong*) received a monthly salted-vegetable pay of 1.2 *liang*, squad leaders (*bazong*) 0.6 *liang*, and 'corporals' (*waiwei*) and common soldiers 0.45 *liang*. There were no daily rations. Officers from Hunan and Hubei, on the contrary, were given their daily rations, but no salted-vegetable pay. Soldiers were given rations and half their salted-vegetable pay (0.4 *liang*). In the beginning of the war, troops from Hubei and Hunan were even given (the money for) 3 *jin* (1.8 kilos) of firewood, a practice which was criticized by the Ministry of Revenue and therefore forbidden later. The numbers of manservants (for officers), or supplementary troops (for common soldiers) in Guizhou and Hu-Guang (Hubei, Hunan) were identical to those of the Sichuan troops, but were lower for the Yunnan troops, for which province the numbers were identical to those in the later regulations of the *Junxu zeli*. For the Yunnan troops, salted-vegetable pay was as high as that for the Sichuan troops.

The paragraphs in the *Jinchuan junxu li'an* dealing with the daily rations regularly specify sums of money allotted to buy rice (1 *fen* or 0.01 *liang* for 0.83 *sheng* [860 ml] of rice from the state granaries, or for 1 *sheng* [1035 ml] of flour).[73] Yet the respective regulations of the *Junxu zeli*, which were compiled on the basis of the precedents from Sichuan (viz. the Jinchuan campaigns), talk of rice, which was handed over to the troops.[74] It might therefore be that the statements in the *Jinchuan junxu li'an* were made to provide a better overview of the cost, which is more easily done by noting down amounts of money rather than volume units of grain or flour. Yet the real value of the rice reaching the camps was much higher than the official conversion rate of 1 *fen* per 0.83 *sheng*, so that the official pay for rations (*kouliang*) is only a theoretical sum.

The regulations for the salted-vegetable pay for Banner troops are not specified enough in the *Jinchuan junxu li'an*. There is a small paragraph about the pay for the provincial Banner troops in Chengdu, and a paragraph describing the general regulations for the salted-vegetable pay for Banner troops deployed to Jinchuan, which is largely based on precedents from the western campaigns against the Dzungars. At least, the listing is

---

[73] *Jinchuan junxu li'an*, juan 1, fol. 25b, 28a.

[74] *Hubu junxu zeli*, juan 2–3.

very detailed for each rank of the Capital Banner troops.[75] The monthly pay for Banner troops—even that for those from the Capital—did not exceed that for the Green Standard troops. The higher cost for the Banner troops therefore originate primarily in the regular higher deployment cost (baggage pay), but then in the irregular presents and rewards given to them as a kind of motivation.[76]

The regulations concerning the daily rations for the native auxiliary troops as reported in the *Jinchuan junxu li'an* differ from those of the later *Junxu zeli*. The most important discrepancy is that the native auxiliary troops, from the kings (acting as native officers) down to common soldiers, were given a daily pay and not the otherwise usual monthly salted-vegetable pay. In the procedures for the 'barbarian affairs' at the beginning of the campaign, each Green Standard soldier was likewise handed a daily salted-vegetable pay of 1 *fen* (corresponding to 0.3 *liang*/month).[77] The monthly pay for the native auxiliaries was, as already noted, not called salted-vegetable pay, but encampment pay (*zuoxiang yin*, if handed out *in specie*, called *zuoxiang mi* 'encampment grain').

In the beginning auxiliary troops were monthly given the money for 3 *dou* of rice (with an official price of 0.255 *liang*), which corresponds to one *sheng* of rice per day, somewhat more than for the other types of troops. Native troops could also be given 1 *sheng* of flour instead of the rice, because the native people were more accustomed to eating *tsampa* (a kind of porridge made of roast barley flour) instead of rice. If the soldiers wanted money instead of food, they were given 8–9 *fen* per *sheng* or rice, according to the old regulations of the transport stations.[78]

Gratuities and Compensations

War is a business in which death and highest honours rub elbows with each other. Promotion, premiums, and presents to high officers might cost the government a lot—medals or other decorations less—and allowances for funerals and pensions for widows likewise stress the budget of a war-waging state. It was therefore also necessary to restrict or at least regulate the cost of these items as well in order to curtail expenditure,

---

[75] *Jinchuan junxu li'an*, juan 2, fol. 25a–27b.
[76] Dai, "Qing State, Merchants", p. 77. Dai, "Military Finance", pp. 303–304.
[77] *Jinchuan junxu li'an*, juan 1, fol. 3a; juan 2, fol. 32b.
[78] *Jinchuan junxu li'an*, juan 2, fol. 33a–33b.

yet without damaging the image of the dynasty as that of a benevolent-patriarchal ruling house.

There are only a few lines of regulations in the *Jinchuan junxu li'an* stating what kind of award and how much of it officers and common soldiers should be given. Yet it was also prescribed which sums being spent for awards could be settled in the accounts. The prescribed regular sums for rewards were given according to monthly pay, for example, "[this and that unit] is to be rewarded with an additional one-month allowance of salted-vegetable pay".[79]

This procedure to reward troops can also be seen from a list of rewards written down in the Jinchuan regulations,[80] where troops were rewarded with one-month extra pay, or an extra salted-vegetable payment of one month. Extra payment could also be granted to individuals, like one month of salted-vegetable pay or half a monthly salary for high officers.

The normal way by which soldiers and officers were rewarded for bravery (apart from being granted money or a promotion), was a nominal rank according to merit (*jungong* 'military merits', in 6 grades). Merits were classified on proposal on the basis of the relation of the own troops' strength to that of the enemy.[81] Yet only with the compilation of the *Junxu zeli* in the late 1770s this process became regularized or even institutionalized at all. Officers could be promoted and were recorded in honorific registers for several generations. Common soldiers were granted a certain amount of money,[82] but could also be given medals (*gongpai*).[83] Everyone could obtain one or several medals of different classes.[84] What neither the *Junxu zeli* nor the Eight Banner statutes (*Baqi tongzhi*) specify is the granting of the Manchu honorific title of 'hero' (*baturu*) that was regularly accompanied with the payment of 100 *liang* of silver.

There were only very rough regulations for rewards. Even after the War Expenditure Code was issued, it remained unclear how common troops or officials could be rewarded. In earlier times, highest commanders had disbursed private funds to reward their troops, yet during the second

[79] *Jinchuan junxu li'an*, juan 2, fol. 1a.
[80] *Jinchuan junxu li'an*, juan 1, fol. 30a–78a.
[81] *Bingbu junxu zeli*, juan 4, fol. 3b–4b.
[82] *Bingbu junxu zeli*, juan 3, fol. 1a–1b.
[83] *Jinchuan dang* 39/I/00324 (QL 39/2/4).
[84] The medals are described in *Baqi tongzhi*, juan 36, pp. 648–651. About the concrete appearance and the fabrication of medals, including cost, see (Jiaqing 22) *Gongbu xuzeng zeli*, in *Gugong zhenben congkan*, vols. 294–298, as *Gongbu zeli san zhong* (Haikou: Hainan chubanshe, 2000), juan 86.

Jinchuan war a large amount of public funds was allotted for the purchase of presents not only for the allied native rulers, but also for the own troops. As Dai Yingcong notes, the Qianlong emperor was "particularly shocked" by the size of rewards made to the troops at the expense of the war chest.[85]

The second type of extraordinary allowances is made up of compensations for soldiers killed in action, as well as for civilian officials and labourers who died from their wounds received in the war theatre or from diseases contracted there. There are many cases where officers survived musket bullet wounds, probably because the Jinchuan rebels did not only use lead for their bullets, but also stones, the latter being possibly less lethal than metal bullets.

The *Jinchuan junxu li'an* distinguishes four categories according to which compensation is given:[86] civilian officials, Banner troops, Green Standard troops, and native auxiliary troops. This is the same pattern as later recorded in the *Junxu zeli* regulations, yet the Jinchuan regulations also talk of a fifth group of people who could suffer wounds or death: the many civilians labouring for the army. For them the *Junxu zeli* does not contain any regulations.[87]

How many people exactly were killed or wounded cannot be known. The introductory chapter of the Jinchuan precedents speaks of 14,731 casualties in total, 908 of whom were civilian and military officials and 13,823 soldiers of all types. People 'suffering wounds' were more than 150,000. Yet this figure also includes the casualties from the campaign against the Muslims in Uš in 1765, the Myanmar campaigns in 1766–1769, and the suppression of the Wang Lun rebellion in Linqing, Shandong, in 1774.[88] In the second part of the Jinchuan regulations, an appended listing of personnel killed contains figures higher than these given in the introductory chapter.[89] According to this part, 686 Banner troops were killed, 15,244 Green Standard troops, and 3,290 native auxiliary soldiers, making a total of 19,220. Five people are listed as having died from some disease. The difference in these figures can be explained by two arguments: the first one, rather speculative, is that the figure of the introductory chapter only refers

85 Dai, "Military Finance", p. 311.

86 *Jinchuan junxu li'an*, juan 2, fol. 1a–5b.

87 *Bingbu junxu zeli*, juan 1, fol. 7b–8a; juan 3–5.

88 *Jinchuan junxu li'an, Zonglüe*, fol. 3a–3b. To this insurgence, compare the book of Susan Naqin, *Shantung Rebellion: The Wang Lun Uprising of 1774* (New Haven: Yale University Press, 1981).

89 *Jinchuan junxu li'an*, juan 2, fol. 5b–7a.

to the Green Standard soldiers killed (without officers), which adds up to 14,793 persons. The second argument is that the introductory chapter only speaks of people killed in action, while the later figures list all people killed in action and those who died later from wounds received during battle. These figures apparently do not include the officers and civilian officials killed during the Mugom uprising (as civilians are not included in this list), which add up to 46 Banner officers and 25 Green Standard officers, plus 3,919 casualties of Green Standard troops and 97 of native auxiliary troops.[90] It is not clear if the last two items are included in the figures of 15,244 and 3,290, but it is quite probable.

All in all a sum of 1,491,312 *liang* for all rewards and compensations was spent, which constitutes only 3 per cent of the total war expenditure.[91]

Civilian Officials

Although war is actually a matter of the military, the troops do not fight for their own purposes, but act on orders of a civilian government. It is therefore necessary that from time to time government representatives have a look at what happens in the war theatre. Even the highest commanders were often no professional generals but governors (*xunfu*) and governors-general of provinces who had to take part in the war because the action took place on or near the territory of the province they were responsible for. The governors-general of Sichuan were therefore automatically involved in the two Jinchuan wars, those of Shaanxi and Gansu in the western campaigns.

Commanding generals often did not act alone but were assigned a civilian official by the central government, often a member of the State Council and thus bearing high responsibility and enjoying the confidence of the emperor. After the Mugom catastrophe, for example, all divisions on the particular routes were commanded by a mixed military-civilian team of officials: Mingliang – Fude, Agui – Sebtengbarǰur, Fengsheng'e – Hailancha. The 'militaries' had the position of a general (*jiangjun*), the 'civilians' that of a Grand Minister Consultant (*canzan dachen*).

Some officials were even dispatched on special order by the emperor to investigate corruption cases, like Fulong'an who was sent to look into the corrupt activities of A'rtai, or Zhou Huang, Baya'r and Fuxing, who were

---

90 *Pingding Liang Jinchuan fanglüe*, juan 66, fol. 9b–10b (QL 38/7/renyin).
91 *Jinchuan junxu li'an*, juan 2, fol. 177a–178b.

separately sent out to find out why the army had failed during the uprising in the summer of 1773.

Governors of other provinces had to raise funds from the local gentry and from among the officialdom within the domain they were responsible for. Probably the most important civilian person involved in the financial organisation of the war was the provincial (financial) administration commissioner (*buzhengshi*) in Chengdu.

The *Jinchuan junxu li'an* as well as the *Junxu zeli* contain long lists of civilian offices in the central government whose occupants could or were to participate in a campaign. Especially officials from the Censorate (*yushitai*), the Grand Secretariat (*neige*), the Hanlin Academy (*hanlinyuan*), the Directorate of Education (*guozijian*), the Directorate of Astronomy (*qintianjian*) and the Imperial Academy of Medicine (*taiyiyuan*) could be dispatched to the front, from ministers and directors down to secretaries, to take over various tasks in the military camps.[92] Among the victims of the Mugom incident two were secretaries in ministerial bureaus (*zhushi*). This could be taken as a hint that the respective ministries, viz. that of Revenue and that of Justice, dispatched secretaries whose task it was to make sure of correct accounting. Representatives of the Ministry of Justice (*xingbu*) might have been dispatched to Jinchuan with the task to check if state officials sentenced to serve in the camps or in the administration, really did their work. A'rtai, for example, had been stripped off his rank of governor-general and was to serve his sentence in the organisation of logistics. This kind of punishment was also known in Europe for generals whose achievements in the field were unsatisfactory.[93] A large amount of civilian officials on the local level were dispatched to organize the logistics network in the hinterland. Such duties were also taken over by persons that had passed the civilian state examination and were awaiting an appointment to a vacant post (*houbu*). The involvement of all these persons into the war happened within the range of normal duties of civilian officials, and they were not given an extraordinary payment for this irregular service. At least they were, under certain conditions, granted the same special allowances like their military counterparts, namely baggage pay, salted-vegetable pay, and daily rations.

---

92 *Jinchuan junxu li'an*, juan 2, fol. 41b–43b. *Hubu junxu zeli*, juan 2, fol. 2a–3a.
93 Van Creveld, *Supplying War*, p. 43.

The *Jinchuan junxu li'an* says that 740 government officials served in Jinchuan,[94] and in another place presents a list stating how many civilian officials served at what route and enumerating 284 people from the rank of prefect down to assistants.[95]

A list of the civilian officials who died during the Mugom disaster will show a typical composition of civilian staff in an important camp.[96] In the encampment of Mugom—at that time the headquarters of the division advancing against the rebels' main strongholds—and during the flight twenty-six civilian officials lost their lives. Three of them were ministerial secretaries (*zhushi*), one a provincial record keeper (*zhaomo*); apart from them there were two prefects (*zhifu*), one sub-prefectural magistrate (*tongzhi*), four department magistrates (*zhizhou*), two assistant sub-prefectural magistrates (*tongpan*), six district magistrates (*zhixian*), one assistant district magistrate (*xiancheng*), one chief of police (*limu*), one police officer (*xunjian*), and three district jailers (*dianshi*).[97] Surprisingly, the lists of casualties do not include any officials from the central government at all, except a few secretaries.

One famous example of a central government official in Jinchuan is the dispatch of Felix da Rocha, vice director of the Directorate of Astronomy (*qintianjian jianfu*), who was to instruct the gunners on the subject of ballistic calculations. For many high-ranking officials their original post is mentioned in the memorials, like that of the ministers (*shangshu*) Agui and Fulong'an or that of the vice ministers (*shilang*) Yuan Shoutong, Liang Guozhi, Guilin, Liu Bingtian, Ebao, Fukang'an, or that of the vice directors (*yuanwailang*) Mingde and Liushiwu.[98] Wenfu's son Yongbao was a reader-in-waiting (*shidu*) in a central government agency, presumably the State Council (*junjichu*).

According to the precedents for vice ministers (*shilang*) and consultants (*canzan*) listed in the *Jinchuan junxu li'an*, a Grand Minister was given a monthly salted-vegetable pay of 12 *liang* and was served on by 24 manservants. Each interpreter (*tongyi*)[99] in his service was given a monthly pay of 1 *liang*. Apparently each manservant was allowed to have one interpreter,

94 *Jinchuan junxu li'an, Zonglüe*, fol. 3b.
95 *Jinchuan junxu li'an*, juan 1, fol. 16a–17b.
96 *Pingding Liang Jinchuan fanglüe*, juan 66, fol. 9b–10b (QL 38/7/renyin).
97 *Jinchuan dang* 38/III/00075-84 (no date).
98 Personal names in the form of numbers were quite common among Manchus.
99 There might have been interpreters for oral communication between the imperial troops and the native population as well as translators for the paperwork of bureaucracy, as the term *yizi* 譯字 suggests (*zi* being something written).

which means that the manservants were not only used to clean their master's shoes and to prepare food, but had much more important tasks to fulfil which required intensive communication with the surroundings. Civilian officials from the Capital (*jingyuan*) were given 30 *liang* monthly and were supported by two secretaries (*shuli*) who obtained a wage of 4 *liang* a month each, further four manservants for whom daily 0.09 *liang* was spent on food. Each secretary was supported by one interpreter.[100] The information in the *Junxu zeli* is more complete: Each Grand Minister was given a baggage pay of two years' salaries, according to his rank, and each manservant 'leather clothing pay' (*piyi yin*) of two *liang*.[101] This regulation followed precedents, which means that Grand Ministers dispatched to Jinchuan should also have received this pay—although there is no mentioning of it in the Jinchuan precedents. For the civilian officials of the local government, there was apparently no baggage pay at all. Concerning the salted-vegetable pay, the *Junxu zeli* quotes also many more precedents and says that all civilian officials of the central government received their monthly pay according to their rank. There was one case in which a high official called Liu Bingtian claimed a monthly pay of 30 *liang* for the travel expenses of one of his subordinates, a certain Lü Yuanliang, because there had been a precedent during the Myanmar campaigns. Yet the payment of the sum Liu claimed was refused by the Ministry of Revenue.[102]

Of much higher significance in pure numbers are the officials of the local government who received the amounts of salted-vegetable pay listed in Table 6.[103] It is very important to compare the figures cited in the regulations for expenditure of the Jinchuan war with the later final regulations in the *Junxu zeli*, because they are much higher than those, just like in the earlier Yunnan precedents, where a prefect (*zhifu*) was granted 30 *liang* of salted-vegetable pay monthly. The *Jinchuan junxu li'an* also quotes the respective amounts for the western campaigns, in the course of which the civilian officials were given an annual instead of a monthly pay, which was substantially higher than in the final regulations of the *Junxu zeli* (360 *liang* instead of 240 *liang*, which became applicable later). There was a discussion in the respective agencies about the salted-vegetable pay for the civilian officials, and it was decided that these should also receive a

100 *Jinchuan junxu li'an*, juan 2, fol. 41b–42a.
101 *Hubu junxu zeli*, juan 1, fol. 1a–1b.
102 *Hubu junxu zeli*, juan 2, fol. 2a–3b.
103 *Jinchuan junxu li'an*, juan 2, fol. 41b–43a. *Hubu junxu zeli*, juan 2, fol. 6a–7a.

pay geared to the fixed annual salary, with the result that the monthly pay was lowered considerably after the second Jinchuan war.

The real cost of the civilian officials cannot be found out because of the fact that in the general account in the *Jinchuan junxu li'an* it is categorized under the heading "Pay for civilian officials, constructing roads and bridges and crafting boats", with a sum of 912,172 *liang*. The sum is composed of the expenditure for 89 different items without separating the particular entries from each other.[104]

Labourers, Porters and Craftsmen

Except civilian state officials, there were a lot of other civilian employees serving in the logistics of campaigns undertaken by the eighteenth-century Qing dynasty. Professionals worked in the logistics stations and camps in various trades like blacksmiths, vets, boatbuilders, casters, physicians, painters, and so on. Other employees oversaw the transport of grain and military items, like accountants and secretaries, or simply foremen. The third group were the least specialized, which only served to do the hard work of transporting rice and ammunition to the camps and to the front.[105]

At the beginning of the war, the local government recruited porters from several districts in Sichuan who were obliged to make available their physical strength as part of the tax they had to pay to the government. The difference between the earlier practice to recruit unpaid *corvée* labourers and the new conscripts for military labour is that the latter were at least given some money for their work. It was not much, but something to live on and it even represented some sort of pay for their way to their destination and back home, and on days when waiting for fresh work to do. Compared with contemporary France, this method shows how much China was commercialized and, from the human point of view, probably more progressive.[106] The recruiting was undertaken according to the tax rate of the region liable for recruitment. For each 100 *liang* of taxes, between 2 and 10 labourers were to be supplied. Persons not wanting to be recruited were allowed to pay a 'voluntary' subsidy. As a compensation to the local community, the *jintie* surcharge (a kind of additional tax used as funding for labour subsidies and the transport of tax grain) was suspended in

[104] *Jinchuan junxu li'an*, juan 2, fol. 177b–178a.
[105] See the article by Dai, "Qing State, Merchants".
[106] Dai, "Qing State, Merchants", p. 80.

districts which had provided labour force to the government. This way, about 150,000 labourers had been recruited from throughout the prefectures and districts of Sichuan.[107] The locations where the labourers came from were spread over the whole territory of the province of Sichuan, so that there was virtually no district where peasants were not to serve the army.

On the long road to the southern route and onwards to Djanggu on the Greater Jinchuan River, for each *dan* of rice two porters had to be used, who were to be paid 0.05 *liang* when working inside the country (*kounei*), and 0.08 *liang* when working outside the country (*kouwai*), but not given any rations, neither inside nor outside the country. Native porters (*fanfu*) working inside the country, i.e. between Yazhou (modern Ya'an) and Dartsedo (modern Kangding), were only paid a daily 0.05 *liang* per person. As the situation for 'barbarian' workers (*manfu*) was the opposite of that for people coming from Sichuan—because the region 'outside the passes' (*kouwai*) is nearer to their homes—they were not even given any pay outside the country but just a daily ration of 1 *sheng* of rice. Porters working at a logistics station were given a daily pay of 0.05 *liang* and obtained a ration of 1 *sheng* of rice. While two porters were to carry one *dan* (103.5 l) of barley the quantity of roasted flour to be carried by two porters was 1.5 *dan* (155.25 l). This difference is due to the bulk density of flour that is smaller than that of rice, leading to a greater volume to be carried when transporting the same weight.[108] Labourers serving on a voluntary basis that were hired on the 'job market' were called 'guest labourers' (*kefu*). They were first made use of during the war against Wu Sangui during the Kangxi reign.[109]

Subordinates of a civilian official (*guanyi* 'official labourers') supervising a logistics station as 'station head' were given a monthly allowance of 12 *liang* and paid 0.03 *liang* for rice,[110] each *fen* being worth 0.83 *sheng* of rice. Civilian runners (*chaiyi*) were monthly paid 8 *liang* and given 2 *fen* to

---

107 *Jinchuan junxu li'an*, juan 2, fol. 71b–72a. Dai, "Qing State, Merchants", pp. 45, 49.

108 The bulk density of polished rice is 0.75 t/m³, that of wheat flour 0.55 t/m³. http://www.bv-net.de/englisch/080_service/08200_bulk.htm. Of course, moisture plays an important role for the weight, so that figures in other sources might differ from these.

109 Dai, "Military Finance", p. 309.

110 3 *fen* of silver bought only about 4 *sheng* of rice. It can therefore be doubted that 3 *fen* was a monthly allowance for rations, and it should have been a daily pay with which the respective persons could buy rice or other food from sutlers, be it for themselves and/or for manservants. A runner was paid 2 *fen*, which was sufficient to purchase rice for two to three persons daily. This assumption is supported by the statements for ration payments for boatmen, see below.

buy food with. People working in other trades, like granary accountants, runners, interpreters and grain measurers, obtained rations of 0.83 *sheng* of rice or 1 *sheng* of flour. They also obtained a monthly pay that was fixed to, according to the notes of the *Junxu zeli*, 1 *liang* monthly.[111]

Boatmen and blacksmiths were paid a family allowance of 1 *liang*, and received 2 *fen* daily for food. When reaching the station where they were due to do their work, they obtained a daily wage of 0.0666 *liang*,[112] and rations of 1 *sheng* of rice.

Two soldiers each were helped by one porter (*beifu*) who was given a wage of 0.05 *liang* daily, but neither rations nor pay when not working or on his way back to his home district (*huikong*).[113]

Yet far more services were provided to the army: Carpenters had to set up the logistics stations and courier stations, stables, storerooms and accommodation, and to build bridges over mountain gorges—sometimes in cooperation with stone cutters.[114] Leather or timber boats had to be built by boatbuilders, blacksmiths had to produce nails and tools; while the Banner units had professional iron smiths of their own, Green Standard units probably had to fall back on civilian professionals. Casters operated the furnaces in which the brass and iron to cast cannon and cannonballs from were molten down,[115] and transporters worked in the batteries bombarding the Jinchuan war-towers.[116] The fuel necessary to operate those furnaces was produced by charcoal burners.[117] Ferrymen and boatmen had to care for a safe crossing of the two Jinchuan rivers and its affluents.[118] Veterinaries and physicians had to cure the wounded and sick.[119] Painters painted scenes and probably drew maps for the emperor and the State Council. Paper makers and artisans mounting pictures or maps produced the necessary material for the painters or map drawers.[120] Tailors repaired tents, made new clothing for the soldiers and probably tailored garments from the brocade brave soldiers were rewarded with. Clerks (*shuli*) from

---

111 *Hubu junxu zeli*, juan 6, fol. 2b–3a.

112 Such a sum is surely the result of a calculation, as for example, 3 *qian* (0.3 *liang*) for 2 persons. *Jinchuan junxu li'an*, juan 1, fol. 5a.

113 *Jinchuan junxu li'an*, juan 1, fol. 3a.

114 *Jinchuan junxu li'an*, juan 2, fol. 69a.

115 *Jinchuan junxu li'an*, juan 2, fol. 67a–67b.

116 *Jinchuan junxu li'an*, juan 2, fol. 68b.

117 *Jinchuan junxu li'an*, juan 2, fol. 69b.

118 *Jinchuan junxu li'an*, juan 2, fol. 69a.

119 *Jinchuan junxu li'an*, juan 2, fol. 69a–69b, 85a. See also *Jinchuan dang* 38/IV/00297 (QL 38/12/13).

120 *Jinchuan junxu li'an*, juan 2, fol. 69a.

the civilian government dealt with the burden of paperwork.[121] And in winter it was even necessary for snow shovelers to clear the mountain paths, as the Jinchuan region was famous for its heavy precipitations.[122] Grain measurers and watchmen for the stations have already been mentioned.[123] For native porters, the so-called *ulaa*, some special regulations were in force.[124]

Yet it is not possible to reconstruct the expenditure for most of those workmen or service providers, due to the lack of figures about the number of persons thus employed. We do not have the slightest idea how many physicians worked in the encampments, and despite the fact that there were clear prescriptions on how many nails each blacksmith had to produce during one working day there is no way to estimate how many of them served in the war theatre. All estimates in this direction can only give a very crude idea of the cost the civilian staff working for the army caused. At least it is known how the payment conditions of these civilians were and to gain an idea of the value of labour in 18th century China, at least from the aspect of how much the government was willing to pay.[125]

All labour cost for workmen (*gongjiang*) had to be brought to account according to the precedents for construction work (*gongcheng zeli*). The first interesting point is that for some labourers and workmen, the family allowance was actually—at least in the beginning of the campaign—not a gratification, but given on loan to be either deducted from the monthly pay or paid back later (in the case of cannon casters and station porters). This kind of procedure is similar to the baggage pay for soldiers, which was originally also only given on loan. The second interesting point is that workmen could also be recruited as labourers (*jiangyi*) and were sent from Sichuan to the war theatre. Such persons probably worked in state-owned workshops supplied with iron, brass and other metals to cast bullets, cannons and cannonballs, and therefore the most logical method was to order state-employed workmen to accompany the metals to the

121 *Jinchuan junxu li'an*, juan 2, fol. 70a.

122 *Jinchuan junxu li'an*, juan 2, fol. 67b.

123 *Jinchuan junxu li'an*, juan 2, fol. 70a–70b, 71a.

124 *Jinchuan junxu li'an*, juan 2, fol. 65b. The Mongolian term *ulaa* (Chinese *wula*) was originally used for the Mongolian postal service as well as the horses used for it, but also for the persons recruited to render the service and the relay stations. See also Peter Olbricht, *Das Postwesen in China unter der Mongolenherrschaft im 13. und 14. Jahrhundert* (Wiesbaden: Harrassowitz, 1954).

125 Data following *Jinchuan junxu li'an*, juan 2, fol. 65a–70b, and *Hubu junxu zeli*, juan 5, 6 and 9.

front. But there were also many workmen simply hired from among the population (*gujiang* 'hired workmen').

The commentary to the *Junxu zeli* is sometimes quite helpful to find out about regulations on items not recorded in the *Jinchuan junxu li'an*, yet there are some smaller items in connection with which the commentary contradicts the figures in the Jinchuan precedents, like the amount of daily rice rations.

The real figure for the expenditure for labourers and workmen can only be established very roughly because in the general account as found in the *Jinchuan junxu li'an* the respective sums are included in different categories, like under the headings "Staff in logistics stations, hiring workmen and family allowances", "Horses and labourers in courier stations", "Transport of grain", "Transport of equipment", and "Pay for civilian officials, constructing roads and bridges and crafting boats".[126] From those figures it is at least possible to see that the staff of the logistics stations, the workmen and the establishment of infrastructure and traffic was not very expensive. The largest amounts of money were disbursed for the transport of grain (making out 71 per cent of the total war expenditure) and of equipment (constituting 8 per cent of the total cost).

Food

The section of the *Jinchuan junxu li'an* talking about the regulations for the purchase and transport of rice, flour and beans says that 2,963,527 *dan* (306,670 m^3) of rice had been bought and transported along the supply lines by both official labourers and commissioned merchants.[127] 2,552,588 *dan* (264,235 m^3) were correctly brought to account as consumed, while an amount of 205,128 *dan* (21,218 m^3) of rice was considered by the Ministry of Revenue as not correctly used, like the rice lost during the Mugom incident and a volume of rice burnt when the grain managers did not take care, but also grain which was not properly dealt with.

The list also states what amounts of rice were transported along which of the four, later five, routes at what point of time (in the old files or the new files), and what amounts of rice were delivered (*shou* 'income') to destinations inside the country (the province of Sichuan) and outside the country. As can be seen, the procedures of accounting for rice were quite similar to those for money, yet it was important to set up separate volume-unit

126 *Jinchuan junxu li'an*, juan 2, fol. 177b–178a.
127 *Jinchuan junxu li'an*, juan 2, fol. 180a–184a.

accounts for rice, flour and beans that were not to be mixed up with the monetary accounts. The importance for separate accounting is also stressed in the *Junxu zeli*, with the argument that unlike money, remaining stock of rice were not allowed to be carried on to the next period of accounting, in order to "avoid confusion".[128] The business of grain distribution among various administrative units and governmental institutions was indeed extraordinarily complex, especially when linking amounts of grain with the cost arising from their acquisition. There must, nevertheless, have been two different accounts for all types of grain, one concerned with the amounts needed, purchased and allocated, and the other with the cost arising from acquisition and transport. The government had quite naturally to deal with a certain price range for grain that kept changing according to time and place. When purchasing grain, nevertheless, the war expenditure bureau had to check with the Ministry whether it would allow exceeding a certain price limit. This was not different from other items to be bought from the market, but unlike with gunpowder or clothes, food was such an essential item that careful planning was necessary in order to prevent troops from going hungry. The quantity of rice thought sufficient to nourish a soldier was 0.83 *sheng* (860 ml) per day, to which other food was to be added, bought with the salted-vegetable pay.[129]

Apart from rice, the troops and labourers consumed 31,902 *dan* (3,300 m^3) of flour and 79,302 *dan* (8,200 m^3) of beans.[130] Beans were needed to feed the horses, while the flour was mainly consumed by the native auxiliary troops and the native labourers in the shape of *tsampa*, and less by the imperial troops to make cakes and rusk with, which was easy to transport and thus an ideal diet on the march, also in the West. Flour was otherwise used as rations when the troops had to spend a period of time without any access to supplies. For this purpose the flour was roasted for conservation.

For the transport of 3,074,731 *dan* of grain (rice, flour and beans) to the logistics stations and the camps the government had spent 30,498,844 *liang*,[131] which means about 10 *liang* of transport cost per *dan*. Unfortunately, this sum is itemized as a total sum for the purchase and transport

128 *Hubu junxu zeli*, juan 9, fol. 10a–10b.

129 *Jinchuan dang* 38/IV/00061 (QL 38/10/10). The emperor asked why 0.83 *sheng* of rice was just sufficient to nourish a Bannerman, while the Green Standard soldiers were not able to consume this amount.

130 *Jinchuan junxu li'an*, juan 2, fol. 180a, 182a, 183a.

131 *Jinchuan junxu li'an*, juan 2, fol. 178a.

of all three types of grain, so that the real price for rice or other types of food has to be looked after in other documents.

The official price of rice in the old files was 1.35 *liang* per *dan*, which included the transport loss surcharge (*jiahao*, amounting to 1–1.5 *sheng* per *dan* and station).[132] Husked rice from donation granaries was not allowed to be brought to account for more than 1 *liang* per *dan*. Between QL 38/8 (September 1773) and QL 39/8 (September 1774) the official price was 1.75 *liang*, yet the Ministry also allowed to bring to account rice with a price of 1.95 *liang*. This regulation remained valid until QL 40/6 (July 1775), yet bad harvests made it necessary for rice to be shipped from Jiading (modern Leshan), Xuzhou (modern Yibin), Chongqing and Luzhou to Chengdu. The price of this rice was calculated according to the officially fixed transport fee for shipping upstream. Arrived in Chengdu, the prices ranged between 2.31 (Jiading rice) and 3.374 *liang* (Chongqing rice). Rice from these prefectures was also transported to the hub Yazhou on the southern logistics route, where it cost between 1.96 (Jiading rice) and 3.024 *liang* (Chongqing rice).[133] These prices were accepted by the Ministry of Revenue as the market prices of rice. The prices for transporting the rice from these two hubs, Chengdu or Guanxian (modern Dujiangyan), and Yazhou, to the camps, were extremely high because the steep mountain range that had to be crossed made it necessary to use the most expensive means of transport, that is, human porters. This problem will be discussed in the chapter about logistics.

The three types of grain were used for special purposes (rice for the Manchus and Chinese, flour for the natives, and beans for the horses), yet there are also cases that they could be interchanged with each other. The daily consumption of rice by porters along the three main logistics routes, for example, was in 1775 more than 700 *dan* (72 m^3). For this reason it was suggested handing out roasted flour every fifth day instead of rice, in order to somewhat cut down the consumption of the latter.[134] Facing the alarming rice prices of 2.4 to 3.6 *liang* in Chengdu, 5 to 6 *liang* in Long'an (Pingwu), Maozhou (modern Maoxian) and Yazhou and up to 10 *liang* in Dajianlu, the logistics managers had indeed every reason to deliberate about a higher amount of the cheaper wheat and barley to be given to the troops instead of the expensive rice.[135] A different argument was brought

132 *Pingding Liang Jinchuan fanglüe*, juan 101, fol. 5b (QL 39/7/jisi).
133 *Jinchuan junxu li'an*, juan 2, fol. 73a–74a.
134 *Jinchuan dang* 40/I/00113 (QL 40/2/8).
135 *Jinchuan dang* 40/III/00139 (QL 40/8/12).

forward in early 1773. Five *sheng* (5.175 l) of rice, it was said, weighed 8 *jin* (4.774 kg) and sufficed to nourish a person for five days. Yet the same amount (8 *jin*) of roasted flour could feed a soldier for eight days.[136]

The price of beans in Sichuan was almost the same as that of rice, because the production of beans in Sichuan was not very high.[137] Due to the price difference being so small and the rice so abundant, at least compared to beans, the horses were often fed more rice than beans, a practice which the Ministry of Revenue criticized as inappropriate. The price for beans was therefore adjusted to 0.7 or 0.8 *liang*,[138] while the official price of rice was kept between 1.35 *liang* and 1.85 *liang*, which is substantially lower than the real market prices.

The government price for flour and wheat grain (*maizi*) was 0.8 *liang* per *dan*.[139] The transport cost for flour was the same, yet the volumes to be carried differed, as already explained earlier. The price of barley (*qingke*), which was almost exclusively consumed by the native auxiliary troops, was 1.85 *liang* per *dan*. The barley was consequently only produced locally and bought in the garrison of Dajianlu.

There are only very few concrete figures concerning the cost of the food that the staff taking part in the second Jinchuan campaign had consumed. The provisions for the secretaries of the officials in the logistics bureaus, which were paid out monthly, had cost 79,444 *liang*. The 2,897 sheep used for food had cost 2,607 *liang* with a price of 0.9 *liang* per animal, yet it is not known if this number of animals had really been bought because each sheep could be substituted by 1.5 *dou* (155 l) of rice, a regulation also included in the *Junxu zeli*.[140] One cow was interchangeable with six sheep, and consequently with 3 *dou* of rice. Sheep meat, along with 'brandy' (*shaojiu*) was often given to the soldiers as a reward or stimulation for future battle.[141] With the salted-vegetable pay the troops could buy additional food, like vegetables or meat, but also tobacco, tea and liquor, in the numerous foodstalls or 'market streets' erected next to the camps. Countless sutlers flocked to these places to offer their merchandise, and there were also money-changers who changed the silver the government

---

136 *Pingjing Liang Jinchuan fanglüe*, juan 85, fol. 7b (QL 39/1/jiwei).

137 *Jinchuan junxu li'an*, juan 2, fol. 85a.

138 *Jinchuan junxu li'an*, juan 2, fol. 74a.

139 *Jinchuan junxu li'an*, juan 2, fol. 74b.

140 *Jinchuan junxu li'an*, juan 2, fol. 57b, 184a, 186b. *Hubu junxu zeli*, juan 8, fol. 1a. The price of a sheep is also given as 0.5 *liang*, idem 2, fol. 85a, or as 0.7 *liang*, *Hubu junxu zeli*, juan 7, fol. 1b. Two sheep corresponded to one monthly ration of meat.

141 *Jinchuan dang* 38/III/00509 (QL 38/9/14); 39/III/00177 (QL 39/8/6).

had given to the troops into copper cash, by means of which the manservants bought vegetables, meat or spices. The emperor was not happy about the sale of liquor and tobacco which he esteemed unnecessary—instead, useful food like sheep should be taken to the camps: He did not want a repetition of a situation like during the western campaigns when the troops had eaten dropped horses, mares and foals when lacking food.[142]

Horses

The ratio between infantrymen and cavalrymen in regular Qing military units was about 3 : 2.[143] Unlike during the western campaigns, where the troops covered large distances in grassland and prairie, it was almost impossible in Jinchuan to make use of cavalry to stage attacks on the enemy.[144] Even the highest Banner officer had to dismount from his 'war horse' (*zhanma*) in the face of the Jinchuan war-towers, and in some places it was—at least at the beginning of the campaign—even prescribed not to allow Bannermen the use of horses in order to save cost.[145] This was not only because the terrain was most unsuitable for cavalrymen, but the custom that officers rode while the common soldiers fought on foot, proved to be dangerous: The horsemen could easily be recognized as officers, and the Jinchuan rebels therefore preferably aimed at the officers.[146] On the march to the war theatre everybody, officer or common soldier, Banner or Green Standard, was entitled to a traveling horse (*xingma*). Yet in one case, when the Banner troops from Jingzhou, Hubei, marched to Sichuan, they abstained from using riding horses (*qima*) and marched on foot to Chengdu, for which reason they were rewarded with a full extra monthly salted-vegetable pay.[147] The troops, of course, had to obtain compensation for this inconvenience, and were given a fixed amount of money, normally 6 *liang*, in exchange for their horses, manservants only 0.2 *liang* (because

---

142 *Jinchuan dang* 39/III/00301-303 (QL 39/9/14).

143 *Pingding Liang Jinchuan fanglüe*, juan 13, fol. 17b–18a (QL 36/12/jichou). This document refers to Green Standard troops. Among the Banner troops there were cavalry as well as infantry units, the latter being more rare. Compare *Baqi tongzhi*, juan 32–35.

144 Lai, *Qianlong zhongyao zhanzheng*, p. 331, has detected one document referring to the use of war horses in combat after it had become evident that in some places the enemy did actually make use of horses for attacking the imperial troops. His assertion that a document in the *Jinchuan dang* 40/III/00069 (QL 40/7/20) informs of three places where the enemy grazed their horses, cannot be substantiated.

145 *Pingding Liang Jinchuan fanglüe*, juan 10, fol. 14a–14b (QL 36/11/bingchen).

146 *Pingding Liang Jinchuan fanglüe*, juan 46, fol. 4a–4b (QL 36/12/gengchen).

147 *Pingding Liang Jinchuan fanglüe*, juan 96, fol. 5a–5b (QL 39/5/wuchen).

they had to share one transport horse) according to the precedents from the Myanmar wars, and 0.1 *liang* during the second Jinchuan campaign.[148]

Horses were extremely important for communication, especially to deliver reports and memorials which had to be sent to Beijing, but also to connect the different routes along which the troops advanced against the enemy. A large number of courier stations were equipped with 'station horses' (*zhanma*), which carried couriers from place to place. The horses for these stations were partly supplied by military garrisons and partly bought on the market.

Huge numbers of transport horses (*tuoma*) were used to take the soldiers' equipment to the front. They were either provided by the home garrisons of the deployed troops (*yingma* 'garrison horses'), or were hired or bought (*caimai ma* 'purchased horses'). In mountainous regions instead of horses often mules (*luo*) were used as beasts of burden. During the western campaigns camels (*luotuo*) served to carry baggage and equipment. The *Junxu zeli* recommends primarily making use of carts to transport military equipment, and only when carts could not be provided, horses or mules had to be made use of. In extremely rough mountains, where the use of horses or carts was out of the question, human porters had to be hired, and as those were the most expensive means of transport, this was the least favoured choice.[149]

In the Jinchuan precedents, information about riding and pack-horses can be found in the chapter about salted-vegetable pay.[150] The horses, just like the baggage pay, had to be paid for by the home garrison, but the money could be advanced from the province, to which it had to be paid back in three-monthly instalments.[151]

Banner officers, it is said, were allowed to ride one horse, which was fed with one bundle of hay inside the country, one bundle being allowed to bring to account with a price of 0.015 *liang*. Outside the country, horses were to be put to pasture and there was no additional allowance for beans or hay. Fodder (*magan*) brought along from the home garrison, could be

[148] *Jinchuan dang* 38/III/00285 (QL 38/7/29). The respective regulation in the *Hubu junxu zeli* 8, fol. 1b–2a, allows each officer to ride a horse (or at least to have one at his disposal) even in mountainous territory. Otherwise, for each horse a compensation of 6 *liang* would be paid. To the compensation for personal assistants, see *Pingding Liang Jinchuan fanglüe*, juan 67, fol. 15b–16a (QL 38/7/wushen).

[149] *Hubu junxu zeli*, juan 5, fol. 1a.

[150] *Jinchuan junxu li'an*, juan 2, fol. 31a–32a, 34b.

[151] *Pingding Liang Jinchuan fanglüe*, juan 40, fol. 12b–13a (QL 37/10/bingzi).

brought to account.[152] The Green Standard troops from Guizhou, bringing along their riding horses, were allowed to feed them 3 *sheng* (3.1 litres) of beans and one bundle of grass per day, weighing 7 *jin* (about 4.2 kilos). Inside the country, this fodder could be bought with a sum of 0.015 *liang*, outside the country with 0.02 *liang*.[153] According to the rules for the courier stations, each horse was to be given 4 *sheng* of rice per day worth 0.04 *liang*. This meant that each horse consumed more than 0.8 *liang* per day.[154] Van Creveld calculates with a relation of 20 pounds (about 9 kilos) of fodder per animal and day, reaped from 80 square metres.[155] John A. Lynn estimates the daily consumption as 17 to 24 pounds of dry fodder, or 50 pounds (23 kg) of grass per animal.[156]

One document in the official history of the second Jinchuan war says that outside the country the Capital Banner officers were given between 1 and 6 horses, depending on their official rank, and between 6 and 20 long-distance porters (*changfu*). Each common Banner soldier was allowed to ride one horse and was supported by three porters. The manservants of both officers and common soldiers were compensated with a pay of 6 *liang* for their horse, which means that they walked.[157] When the Banner troops from the Capital arrived in Chengdu they were paid 15 *liang* from the 'horse fund' (*majia yin*) provided for such occasions to buy or to hire horses or mules to transport their equipment.[158] The Jinchuan precedents as recorded in the *Jinchuan junxu li'an*, as well as later the *Junxu zeli*, give detailed instructions as to how many riding horses and camels a Banner officer was allowed to have at his disposal. Yet these instructions are derived from the western campaigns, during which it was possible to make use of camels. Where camels could not be used or were not available, one camel could be replaced by two horses. In the beginning of the Jinchuan campaign, the character 駝 'camel' (read *tuó*) in the precedence documents was interpreted as the homophonous 馱 'transport horse', for which reason the number of animals was not sufficient (only one horse instead of two horses replacing one camel). The same problem occurred

152 *Jinchuan junxu li'an*, juan 2, fol. 25b.
153 *Jinchuan junxu li'an*, juan 2, fol. 29b.
154 *Pingding Liang Jinchuan fanglüe*, juan 120, fol. 21b–22a (QL 40/7/dingsi).
155 Van Creveld, *Supplying War*, pp. 24, 34, says that one horse needs the fiftieth of an acre per day, i.e. 41 m^2.
156 Lynn, *Feeding Mars*, p. 141.
157 *Pingding Liang Jinchuan fanglüe*, juan 67, fol. 4a–14b (QL 38/7/wushen).
158 *Pingding Liang Jinchuan fanglüe*, juan 68, fol. 5b–6b (QL 38/7/renzi).

for the transport of silver, when horses were replaced by mules, which are not able to carry the same load.[159]

Green Standard officers were, according to the later rules of the *Junxu zeli*, not given any additional transport horses. Soldiers were given one transport horse for two persons and manservants one horse for three persons. Each horse was equipped with a pack-saddle, for which only 0.65 *liang* could be brought to account. In the precipitous terrain of Jinchuan, where it was not possible to use horses, porters had to carry the luggage, with 80 people carrying the equipment of 100 soldiers. If there were supplementary soldiers (*yuding*) in the unit, only 50 porters could be hired, and the supplementary troops had to carry the baggage. The horses of the manservants were to be replaced with a ratio of three porters for two horses, or one porter for two manservants.[160] These proportions are of great help to assess the weight of the load which the horses had to carry: As is known, one porter could be loaded with up to 50 *jin* (29.8 kilos), which means that one horse did not carry more than 75 *jin* (44.8 kilos) in normal terrain, in other words, 1.5 or 1.6 people carried the same load as one horse or mule. One soldier on average had to carry a load of thirty to forty *jin* (about 18 to 24 kilos), of which gunpowder, 200 bullets, fuses, his musket, and dagger already counted for more than 30 *jin*, plus rations for ten days which was itself more than 10 *jin* (each day one *sheng* of rice, which corresponds to 1035 ml, with a total weight of about 9.5 kilos).[161] When adding tent and cooking woks to that weight, the burden for each individual soldier was probably more than 30 kilos.

According to the newly issued regulations of the *Junxu zeli*, native auxiliary troops were only allowed to bring to account riding and transport horses when dispatched to another province, which was not the case in the Jinchuan war.[162]

For the mules rented from Shaanxi the *Jinchuan junxu li'an* contains special regulations. The animals were daily to be given 10 *jin* of hay, one Capital *sheng* (*jingsheng*) of beans and two Capital *sheng* of bran (*maifu*), which was all to be provided from the day the trek left its home district. As long as it was still in Shaanxi, this province had to advance the expense.

159 *Jinchuan junxu li'an*, juan 2, fol. 31b–32a.
160 *Jinchuan junxu li'an*, juan 2, fol. 31a.
161 *Pingding Liang Jinchuan fanglüe*, juan 85, fol. 7b–8a (QL 39/1/jiwei); juan 86, fol. 20a–20b (QL 39/1/renshen). It might be remembered that 5 *sheng* (5.2 litres) of rice weighed 8 *jin* (4.8 kilos).
162 *Hubu junxu zeli*, juan 4, fol. 9a.

When entering the territory of Sichuan, the latter province advanced the cost for the fodder which was now slightly changed: In Sichuan each animal was daily given 2 *sheng* of beans and 2 *sheng* of bran. The government had prescribed by what prices the fodder had to be brought to account: 0.75 *liang* for one *dan* of beans, but no more than 0.8 *liang*, while the fixed price for bran was 0.3 to 0.4 *liang* per *dan*. One bundle of hay could cost no more than 0.0015 *liang*. The animals had to be cared for in order to keep them in good condition, and therefore in each logistics station the mules were to be kept in a stable over night.[163]

The concrete expenditure for all types of horses during the second Jinchuan war is not known. The cost for horses in the courier stations is subsumed under one heading with the cost for labourers in these stations with a total amount of 468,723 *liang*.[164] It is also known that 2,535 horses were bought for the transport along the logistics lines, and that 1,500 garrison horses were lend out for the same purpose.[165] These horses were purchased in order to replace those animals that had died by the hard work in the mountainous region. It was prescribed that no more than 30 per cent of the horses per year dropped inside provincial territory, outside territory a mortality rate of 40 per cent was allowed. The state would pay for these horses, but only under the condition that the remainings of the horses ("skin and bowels") would be sold, for no less than 0.5 *liang*, in order to cover part of the official price for a new horse, which was 8 *liang*. During the first Jinchuan campaign the actual market price of a horse had been 12 *liang*, in Yunnan, during the Myanmar campaigns, the price had been as high as 18 *liang*, and in Shaanxi and Gansu, the average price for a horse was 8 to 10 *liang*.[166] The regulation to sell part of the horse's corpse on the market is very old and is already to be found in the administrative law of the Qin period (221–206 BCE).[167] The courier horses were so important that it was even allowed that in each sixth courier station a veterinary was working.[168]

[163] *Jinchuan junxu li'an*, juan 2, fol. 80a–81a.
[164] *Jinchuan junxu li'an*, juan 2, fol. 178b.
[165] *Jinchuan junxu li'an*, juan 2, fol. 84a, 186a.
[166] *Jinchuan junxu li'an*, juan 2, fol. 84a, 85b–86a.
[167] *Shuihudi Qinjian zhujian*, ed. by Shuihudi Qinmu zhujian zhengli xiaozu (Beijing: Wenwu chubanshe, 1990), *Qinlü shiba zhong, Jiuyuan lü*, p. 33.
[168] *Jinchuan junxu li'an*, juan 2, fol. 85a.

Weapons and Military Equipment

The standard gun of the Qing troops was the so-called 'bird musket' or 'fowling piece' (*niaoqiang*). The name might be derived from the bird-shaped appearance of the serpent holding the fuse (hence the old designation *niaozuichong* 'bird beak musket'). The Chinese muskets were derived from a Central Asian type of musket, as Needham has shown.[169] During the second Jinchuan war 3,574 *niaoqiang* muskets were paid for by the government.[170] Since muskets were considered to be durable equipment and possibly had to be cared for by the user for decades or even generations, these weapons might either have been replacement for broken ones or arms distributed among troops earlier not equipped with such.

Lead (*heiqian*) bullets were, as the *Junxu zeli* says, made and kept ready in the garrisons, where labourers were charged with the task to produce the ammunition. In the camps it might likewise have been manservants who cast the bullets. According to the precedents in Yunnan labour cost was to be paid to the casters, the "labourers of the workshop", the people smoothing the bullets, and for the charcoal burners. The Yunnan garrisons made use of pure or refined lead (*jingqian*), yet the Sichuan garrisons used crude lead (*maoqian*), according to the *Junxu zeli*, which contradicts the rules recorded in the *Jinchuan junxu li'an*.[171] This might be due to the fact that troops from many different provinces took part in the campaign and that for the account of war expenditure a standard between the different provincial rules had yet to be elaborated.

According to the final account book in the *Jinchuan junxu li'an* the generals had used 4,271,475 *jin* (2,550 tons) of gunpowder.[172] According to the price list as presented in the *Jinchuan junxu li'an* it was possible

169 Joseph Needham, *Science and Civilization in China*, vol. 5, *Chemistry and Chemical Technology*, part 7, *Military Technology; the Gunpowder Epic* (Cambridge: Cambridge University Press, 1962), especially pages 440 ff., where the 'unexpected turn' towards the Islamic world is made.

170 *Jinchuan junxu li'an*, juan 2, fol. 185a. Yang Dongliang and Zhang Hao, *Zhongguo Qingdai junshi shi* (Beijing: Renmin chubanshe: 1994), p. 205, say that there were about 270,000 muskets used empire-wide, yet without giving a reference. The private manufacture of firearms was prohibited, compare Dai, "Military Finance", p. 306.

171 *Hubu junxu zeli*, juan 1, fol. 2b–3b. *Jinchuan junxu li'an*, juan 2, fol. 45a–45b. Additional information on the production of bullets can also be found in (Qianlong 58) *Gongbu zeli*, in *Zhongguo gudai jiangzuo ziliao congkan*, as *Gongbu zeli zheng-xubian* (Beijing: Bejing tushuguan chubanshe, 1997), juan 26, and (Jiaqing 20) *Gongbu zeli*, in *Gugong zhenben congkan*, vols. 294–298, as *Gongbu zeli san zhong* (Haikou: Hainan chubanshe, 2000), juan 32.

172 *Jinchuan junxu li'an, Zonglüe*, fol. 3b; juan 2, fol. 184a.

to manufacture 100 *jin* (60 kilos) of gunpowder at a price of 3.05 *liang*.[173] Prices for nitre (*huoxiao*) which was to be stored in the logistics stations range between 3.19 *liang* and 5 *liang* per *dan*,[174] which means that the market prices for nitre varied in a wide amplitude. There were also differences in the price of nitre among the provinces, for which reason the *Junxu zeli* did not fix an empire-wide official price. The ingredients for 'cooking' gunpowder were also different, depending on the climatic conditions. The Ministry of Works fixed as a rule of thumb a ratio of 11.8 *liang* (here: weight measure, about 450 gr) of nitre, 2.1 *liang* of sulphur (about 75 gr) and 2.1 *liang* of willow charcoal per *jin* (or 16 *liang*) of gunpowder. Labour cost was not to be more than 0.0144 *liang* per *jin* of gunpowder. The price of 1 *jin* of nitre was to be no more than 0.0325 *liang*, that for sulphur 0.05 *liang*. While the Yunnan garrisons used pure nitre (*jingxiao*) and pure sulphur (*jinghuang*), the Sichuan garrisons only used 'crude' sulphur and nitre.[175] This statement can be supported by a document reporting that pure nitre and sulphur should be bought in Shaanxi because no pure raw material was available in Sichuan.[176] Sulphur was often bought in Chongqing, for which a transport price of 0.626 *liang* per 100 *jin* was allowed to bring to account, as well as 0.06 *liang* of travel cost per day for an official (officer), and 0.04 for a common soldier.[177] The precedents from the second Jinchuan war were the only regulations made for the production of gunpowder until the compilation of the *Junxu zeli* took place.

The traditional weapon of the Banner troops was bow and arrow. The most widespread type of arrow was the so-called 'plum-head arrow' (*meizhenjian*, written with the character 梅), also known as 'eyebrow-head arrow' (*meizhenjian*, written with 眉), a name derived from the slim shape of the arrowhead resembling an eyebrow. During the second Jinchuan war the Banner units made use of 20,000 *meizhen* arrows.[178] Normal arrows were also used and in significantly higher quantities than this type of elite arrow: 478,500 in all. Although the manufacturing requirements are described in the precedents of the Ministry of Works, no price

173 *Jinchuan junxu li'an*, juan 2, fol. 45a. Regulations for the production of gunpowder can also be found in (Qianlong 58) *Gongbu zeli*, juan 23–26, and (Jiaqing 20) *Gongbu zeli* juan 29–32.

174 *Jinchuan junxu li'an*, juan 2, fol. 62a.

175 *Hubu junxu zeli*, juan 1, fol. 1a–2b.

176 *Pingding Liang Jinchuan fanglüe*, juan 66, fol. 21a–21b (QL 38/7/guimao).

177 *Jinchuan junxu li'an*, juan 2, fol. 45a.

178 *Jinchuan junxu li'an*, juan 2, fol. 185a.

can be determined from this source.[179] Yet the 1804 edition of the *Junqi zeli* "Regulations for military equipment" gives a price of 0.035 *liang* for a normal arrow.[180]

The Qing troops, especially the Green Standard units, used a lot of antiquated weapons the exact figure of which is only known for the short swords (*yaodao*), namely 5,420 pieces,[181] which were presumably additionally produced during the war. Shields of wood or reed (*mupai, tengpai*), spears (*changqiang*), lances (*changmao*) and other polearms are rarely mentioned because they played only a role in close combat of which there are no reports in the official documents. The only exception are 'spectacular' weapons like hand grenades (*huodan*) and 'flame-throwers' (*pentong*). Concerning the hand grenades, there is a detailed description of the fabrication in the *Junxu zeli*.[182] According to this source, there existed no regulations for the production of hand grenades, except in Sichuan. Rockets (*huojian*) were also used, although they are only documented once.[183]

Cannon (*pao*) was not a predominant weapon of the Qing armies. They nevertheless played an important role already in the earliest fights against the Ming armies, as Nicola di Cosmo has shown.[184] There was a specialized Firearms Brigade (*huoqi ying*) among the Banner troops in Beijing, and each Banner garrison disposed of one or several cannons.[185] Yet these guns were mainly used for defence or siege purposes and could hardly be taken into the field. This circumstance did not change greatly during the late 17th and the 18th centuries, as Chinese cannon lacked a carriage with large wheels. Larger distances could therefore only be covered on the back of camels or horses. Furthermore, these beasts were only able to carry smaller pieces that were not heavy enough to blow a breach into a city wall. The solution for this problem was to cast cannon near the war

179 (Qianlong 14) *Gongbu zeli*, also called *Gongbu zeli zheng-xubian*, in *Zhongguo gudai jiangzuo ziliao congkan* (Beijing: Beijing tushuguan chubanshe, 1997), *juan* 45–46.

180 (Jiaqing 9) *Junqi zeli*, in *Gugong zhenben congkan*, vol. 293 (Haikou: Hainan chubanshe, 2000), juan 2, fol. 10a.

181 *Jinchuan junxu li'an*, juan 2, fol. 185a.

182 *Gongbu junxu zeli*, juan 1, fol. 7a–8a.

183 *Pingding Liang Jinchuan fanglüe*, juan 97, fol. 10b–11a (QL 39/11/gengyin). About the production of rockets, see (Jiaqing 20) *Gongbu zeli*, juan 33, and *Gongbu junxu zeli*, juan 1, fol. 6a.

184 Di Cosmo, "Did Guns Matter?".

185 See, for instance, the listings of the fixed amount (*ding'e*) of Banner troops in the various garrisons in *Baqi tongzhi*, juan 32–35. Yang and Zhang, *Qingdai junshi shi*, p. 205, say that there were about 160,000 (small) cannons used empire-wide, yet without providing a reference.

theatre. Already during the western campaigns, cannons had been cast in Gansu and were then taken to the battlefields in the far west. During the Myanmar campaigns, and especially during both Jinchuan campaigns, cannons and howitzers were produced in so-called 'mobile field foundries' (*suiying paoju*). For these cases, special regulations were developed concerning the production, transport and accounting matters of cannons cast near the battlefield. Thanks to these regulations, ample information about the contemporary production of cannons has survived.[186]

Only a few types of cannons or howitzers (all types of heavy firearms were collectively called *pao* in Chinese) are mentioned in the sources. Most relevant documents speak of 'large cannons' (*da pao*), sometimes indicating the weight or the calibre, like "a large brass cannon weighing 3,000 *jin* (1,791 kilos)", "casting large cannons weighing 3,000 to 4,000 *jin*", "small cannons of only 6–700 (or 7–800) *jin* of weight (700 *jin* corresponding to 418 kilos)", so that "several hundred pieces would be necessary to destroy a war-tower and to make it ready to be attacked",[187] "casting a large cannon of more than 5,000 *jin* of weight (2,985 kilos)", "cannonballs made of crude iron with a weight of 20 *jin* (12 kilos) were made", or "the 16-*jin*-calibre (9.6 kilos) cannons being cast". Other cannons were simply labelled by the material they were made of, namely iron (*tiepao*) or brass (*tongpao*). As most cannons were made of brass, the term 'brass cannon' only occurs very rarely, to distinguish them from iron cannons or other types of cannon. Another standard cannon was the so-called 'mountain-breaking howitzer' (*pishan pao*), detailed descriptions of which are to be found in the Jinchuan regulations.[188] There existed also a smaller version, the *pishan xiaopao*.[189] The most important cannons by far were the not fully standardized types which were cast near the camps according to need. One type of cannon that had not to be cast in the camp foundry was the 'Heaven-assailing howitzer' (*chongtian pao*), also called 'water-melon howitzer' (*xigua pao*), which was exclusively used by the elite Banner units and for which special ammunition was brought from Beijing.[190] It weighed between 300 and 380 *jin* (179–227 kilos).[191] The transport of

[186] *Jinchuan junxu li'an*, juan 2, fol. 58b–59a, 63a–64b.
[187] *Pingding Liang Jinchuan fanglüe*, juan 16, fol. 14a–14b (QL 37/1/xinhai).
[188] *Jinchuan junxu li'an*, juan 2, fol. 58b–59b.
[189] *Pingding Liang Jinchuan fanglüe*, juan 58, fol. 16a (QL 38/4/dingyou).
[190] *Jinchuan junxu li'an*, juan 2, fol. 185a.
[191] *Pingding Liang Jinchuan fanglüe*, juan 98, fol. 9b (QL 39/6/guimao); juan 104, fol. 21b–22b (QL 39/9/dingsi).

moulds from Beijing had been organized by Amida.[192] There were also some *chongtian* howitzers used in Jinchuan, formerly transported to Yunnan for the Myanmar campaigns.[193]

Metal that is cast and recast several times will of course lose much of its original quality, and in order to regain the original pureness of the brass, the foreman had to add 8 or 9 *jin* of 'pure' brass to every 100 *jin* of remelted brass. If the brass had already been remelted and recast several times, or had, as during the catastrophe of Mugom, been buried in order to hide it from the enemy and was therefore heavily corroded, the foundry master had to add up to 16.375 *jin* (16 *jin* 6 *liang*) to every 100 *jin* of the corroded material.[194]

In the final account book the officials outline the exact numbers of large brass and iron cannons used during the war: 4 *tiancheng jiangjun pao* 'cannons of the general of celestial completion'; on the western route: 1 'great general cannon' (*dajiang pao*), 7 large iron cannons, 2 'nine-segment cannons', 98 'mountain-breaking cannons', 161 brass cannons 'cast in the camp foundries'; on the central route: 2 iron cannons; on the southern route: 3 iron and brass cannons, 1 small brass cannon, 41 'mountain-breaking cannons', and 117 iron cannons 'cast in the camp foundries'. Although those figures make for a total of only 437, the *Jinchuan junxu li'an* speaks of 657 large cannons.[195]

For all metal items needed in the camps 7,245,461 *jin* (4,325 metric tons) of 'crude and refined brass and iron' (*sheng-shu tong-tie*) were used.[196] Of these, 6,865,170 *jin* (4,099 t) were iron and 391,437 *jin* (234 t) brass. Although it is not possible to attribute the metals needed to particular metal items, like woks, hooks, nails, cannons, and so on, it is possible to calculate the pure material value of the metals. The almost 7 million *jin* of iron thus cost the Qing state about 70,000 *liang*, with an average price of 0.01 *liang* per *jin* of impure iron (*maotie*), or 435,000 *jin* for refined iron (*shutie*) with a price of 0.06 *liang* per *jin*. Prices for iron varied considerably, depending on the local market situation. The Ministry of Works therefore did not fix prices for settling accounts on iron. Nevertheless the bureaucratic framework made it necessary for officials not to invent astronomical prices, and therefore rough directives for iron prices can be found in the *Jinchuan*

192 *Jinchuan dang* 39/III/00177 (QL 39/8/6).
193 *Jinchuan dang* 41/I/00283 (QL 41/3/9).
194 *Jinchuan junxu li'an*, juan 2, fol. 44b.
195 *Jinchuan junxu li'an*, *Zonglüe*, fol. 3b; juan 2, fol. 185a.
196 *Jinchuan junxu li'an*, juan 2, fol. 185b.

junxu li'an, crude iron (*shengtie*) costing 0.008–0.015 *liang* per *jin*, scrap iron (*huangtie*) 0.005–0.015 *liang* per *jin*, impure iron (*maotie*) 0.01 *liang* per *jin*, and refined iron 0.03–0.06 *liang* per *jin*.[197]

Concerning the brass, all calculations have to be based on other sources than the *Jinchuan junxu li'an*: The *Gongbu junqi zeli* "Rules for military equipment by the Ministry of Works" cites prices of 0.13 *liang* for crude brass (*shengtong*), 0.2 *liang* for 'red refined brass' (*hong shutong*) and 0.15 *liang* for 'yellow refined brass' (*huang shutong*).[198] While the former seems to have had a higher content of copper in the alloy, thus being what is called bronze, the latter seems to be more appropriate to what is brass. The price for objects made of 'crude brass' was thus, as can be seen from the data above, more than ten times of such made of 'crude iron'. The process of casting iron cannons was apparently quite simple compared to that of brass cannons, which might be due to the fact that brass cannons were a more recent development and thus technologically more advanced. This can be seen in the large number of different work to be done with the barrel being cast ready, from polishing to boring and decorating, leading to a price of 16.5 *liang* per 100 *jin* for a high-quality brass cannon.[199] For comparison, data on material used to produce iron cannons can be used to roughly estimate a price of 3.2 *liang* per 100 *jin*,[200] while the production of cannonballs can quite exactly be calculated as 2.2 *liang* per 100 *jin*.[201] According to the final account, cannonballs made of crude or refined iron were produced to an amount of 536,546 *jin*.[202] In Europe it was common during the eighteenth century that an army took along 100 cannonballs for each barrel for a siege. Each gun fired no more than four rounds a day.[203]

For the melting process three people from the logistics stations were hired, during the melting and first founding one workman who was no longer needed for the rest of the process. Eight more persons were used to collect fuel and to erect a charcoal kiln. Two charcoal burners observed the carbonisation process. The kiln was opened after five days, yielding

---

197 *Jinchuan junxu li'an*, juan 2, fol. 46a–46b, 64a–64b.

198 *Gongbu junqi zeli*, comp. by Dong Gao et al. (print from 1812; electronic version [data bank] used via licence by Staatsbibliothek Berlin of Zhongguo jiben guji ku, ed. by Beijing Airusheng shuzihua jishu yanjiu zhongxin, Beijing), juan 56, fol. 7a.

199 *Gongbu junqi zeli*, juan 25, fol. 8a–10a, with material costs from *Gongbu junqi zeli*, juan 56.

200 *Gongbu junqi zeli*, juan 25, fol. 7a–8a, with material costs analogous to data in *Jinchuan junxu li'an*.

201 *Jinchuan junxu li'an*, juan 2, fol. 45b.

202 *Jinchuan junxu li'an*, juan 2, fol. 185a.

203 Van Creveld, *Suppyling War*, pp. 35–36.

450 to 460 *jin* (287–275 kilos) of charcoal. Regulations for charcoal kilns which were derived from the precedents from the second Jinchuan war are also included in the *Junxu zeli*. This source states that for an amount of 450 to 460 *jin* of charcoal several thousand *jin* of wood were necessary.[204] Because the natural conditions and the available types of wood for fuel were different in each place, no concrete relation is recorded. The large staff needed for each kiln is justified by the large amount of wood to be collected and the important task to diligently observe the burning process around the clock. The regulations explicitly say that it was only allowed to bring cost for charcoal kilns to account if it was more expensive to provide mined coal from territory inside the provinces, than to burn charcoal on the spot.

The cost for a 'fuse disc' (*huosheng pan*), from which cords were cut down according to need, were allowed to settle for 0.0072 *liang*.[205] Another source, the *Junxu zeli*, gives information that in Sichuan also longer (2 *zhang*) and thinner (2 *fen*) fuses were produced that could be settled for 0.00455 *liang*. Fuses with a thickness of 2.5 *fen* had a government price of 0.005275 *liang*, and those with a thickness of 3 *fen* a price of 0.006 *liang*. The Yunnanese fuses were much cheaper than those from Sichuan because they were not made of flax but from the bark of a kind of tree (*langshu pi*).[206] Although the Ministry did not feel very comfortable with the fact that the material prices were different in each province according to local conditions and therefore rejected the account for a certain amount of fuses, it did in the end only cancel somewhat over 3 *liang* from a sum of 231.37 *liang* brought to the respective account, spent for the production of 11,749,023 fuse discs.[207]

Living and Mobility

As could be seen in the foregoing sub-chapter the government did not only regulate what weapons were allowed to be kept as standard military equipment and how much their production was allowed to cost, but it also determined what types of equipment for living were to be used and were allowed to be brought to account. While tools like hammers, axes,

---

204 *Gongbu junxu zeli*, juan 1, fol. 4a–5a.

205 *Jinchuan junxu li'an*, juan 2, fol. 45b–46a. Additional information about the production of fuses can be found in (Qianlong 58) *Gongbu zeli*, juan 26, and (Jiaqing 22) *Gongbu zeli*, juan 32–33.

206 *Gongbu junxu zeli*, juan 1, 3b–4a.

207 *Jinchuan junxu li'an*, juan 2, fol. 184b.

spades and so on were in most cases included in the regulations for equipment directly belonging to artillery or to various labour teams like sappers and miners, there are three other fields of great importance for the physical and mental well-being of the troops, namely cooking utensils, tents, and clothing. The interesting aspect of those items is that the regulations for war expenditure, the *Junxu zeli*, cite the precedents from Yunnan and Sichuan, i.e. the Myanmar and Jinchuan campaigns, as rules valid for the future.[208] The precedents declare that all equipment, from weapons to tents, woks, and tools, were to be produced according to the rules for the manufacture of military equipment valid throughout the empire and issued by the Ministry of Works. The cost, nevertheless, had to be brought to account according to the precedents for the province concerned. The 1812 edition of the *Gongbu junqi zeli* is therefore divided into two parts, the first part (Chapters 1–42) regulating the composition and production of armour, clothing, tents, weapons and flags, and the second part regulating the question of particular expenses in the various provinces (Chapters 43–60). A very interesting point is that uniforms, clothes and boots were traditionally not provided by the government during a campaign but were to be made and paid by the garrisons and the soldiers themselves. It was therefore not allowed to declare them as war expenditure. The troops of Hubei, for example, annually sent a quarter of their pay to Jinchuan in settlement of their clothes and boots.[209] Tents likewise, as durable equipment, had a government-fixed lifetime and could therefore not be declared as war expenditure. Yet in some cases the emperor granted an exception from this rule, for which reason the *Jinchuan junxu li'an* also contains a number of prices and exact rules for the dimensions of tents.[210] These were nevertheless reduced to two standard types during the war. Tents were only allowed to be used by imperial troops and not by native auxiliaries or military labourers. Only in very exceptional cases without an infrastructure behind a quickly advancing front the emperor allowed that kind of staff to use imperial tents.[211]

The war expenditure regulations from the Jinchuan war are extremely detailed concerning cooking utensils,[212] tents and Mongolian yurts of all kinds and equipped with different linings, even down to single parts as

---

208 *Gongbu junxu zeli*, juan 1, fol. 5a–5b.
209 *Jinchuan dang* 39/III/00233 (QL 39/8/28), 39/III/00317 (QL 39/9/21).
210 *Jinchuan junxu li'an*, juan 2, fol. 50b–58a, 62a–62b.
211 *Jinchuan junxu li'an*, juan 2, vol. 148b–149b.
212 *Jinchuan junxu li'an*, juan 2, fol. 48a–49b, 57a–57b, 64b.

tent pegs and adornments. They list the prices for clothing, coats, caps, tunics,[213] various brocades and satins that served as presents and rewards for native chieftains,[214] medicine,[215] stationery[216] and transport tools like sacks, baskets, palm fibre mats and wooden boxes.[217] It is prescribed how much the construction of a granary, a makeshift bridge, a round leather boat, wooden palisades, or clock towers in a camp were allowed to cost.[218] This price might include the cost of material and labour, yet especially concerning wood, it was prescribed that it was not allowed to bring material cost to account. This is similar to hay for fodder that was, cheap as it might be, not allowed to be declared as war expenditure outside provincial territory.

All these minutiae cannot be listed here in detail, but it is important to know that they have survived and were part of regulations valid only for that particular campaign or which might serve as precedents for a campaign in the same region or a war fought under similar conditions. It would therefore be interesting to know whether these regulations were also valid during the campaigns against the Gurkhas that had invaded Lhasa in 1791. The flexibility with which the Qing government approached this type of war expenditure demonstrates that all matters concerning state employees could be regulated centrally, which is also true for transport conditions and transport cost, while the cost of material, including the important factor rice, was always geared to real market conditions in the region concerned. Yet this does not mean that the government was willing to pay all market prices, otherwise no regulations would have been formulated.

The cost for constructing bridges, boats and roads during the second Jinchuan war had accrued to 4,850 *liang*, which is surprisingly little for a four year long war that made necessary the setup of a large logistics network in inaccessible territory. It is also known that 126 boats and rafts had been constructed during the war.[219] The concrete expenditure for all equipment might be included in the sum of 912,172 *liang* spent as "Pay for civilian officials, constructing roads and bridges and crafting boats, etc."

---

213 *Jinchuan junxu li'an*, juan 2, fol. 47a–48a.
214 *Jinchuan junxu li'an*, juan 2, fol. 60a.
215 *Jinchuan junxu li'an*, juan 2, fol. 139b–140a.
216 *Jinchuan junxu li'an*, juan 2, fol. 140a.
217 *Jinchuan junxu li'an*, juan 2, fol. 44b–45a, 46a, 60a–61b.
218 *Jinchuan junxu li'an*, juan 2, fol. 71b, 88a–89a, 90a, 186a.
219 *Jinchuan junxu li'an*, juan 2, fol. 186a–186b.

The Auditing Process of a War Expenditure Account

At the beginning of a campaign the responsible governor-general set up one or more war logistics bureaus (*junxuju*) and expenditure bureaus (*jingfeiju*) in the provincial capital or other strategically favourable places.[220] These bureaus were run by several high officials that had in hand the respective regulations for war expenditure. The regulations helped them to gear their expenditure towards a fixed official price. Accounts using the prices stated in the expenditure regulations would be accepted by the Ministry of Revenue, all others had to be discussed under certain conditions or were not accepted. Any surplus expenditure had to be covered by other means. In most cases it can be assumed that expenditure was then claimed for items or services that were not used in fact, the money obtained for which was then used to cover overspending. In the worst case, the responsible officials had to cover surplus expenses with their own salary. This method helped the central government to cut cost as far as possible. The *junxuju* is an invention by the Qing and was born in the early eighteenth century out of the need to have a better control on finances, as a result of the proposition that military expenditure had to be covered by the government solely.[221]

The war expenditure bureaus were responsible for the organisation and pay of the logistics and the wartime pay for troops and civilian personnel. The number of officials, secretaries, assignees, scribes and copyists working in the war expenditure bureau was not fixed, but they were allowed to bring to account the cost for items like stationery or lamp oil.[222] When the war was over, an accounting bureau (*baoxiaoju*) was set up to check all accounts and prepare them for final submission to the Ministry of Revenue. The accounting bureau was divided according to cost groups, from troops and horses to grain and military equipment. As in many central government institutions there was also a department exclusively occupied with producing fair copies of the documents to be submitted. The staff in the accounting bureau was largely the same as in the former war

---

[220] There is only one reference to the expenditure bureau in *Jinchuan junxu li'an*, juan 2, fol. 140b, and scarce references elsewhere to this kind of institution. Therefore in this book the logistics and the expenditure bureaus will be regarded as a single operating unit.

[221] The first official references to the *junxuju* are mentioned in 1757, *Pingding Zhungar fanglüe, Zhengbian*, juan 37, fol. 22a (QL 22/3/yiwei), and 1759, *Gaozong shilu*, juan 593, fol. 37b (QL 24/7/dingchou). The institution itself might be somewhat older than those dates.

[222] *Jinchuan junxu li'an*, juan 2, fol. 140a, 186b. *Hubu junxu zeli*, juan 7, fol. 1b–2a. Lai, *Qianlong zhongyao zhanzheng*, p. 384.

expenditure bureau, so that individual expertise and experience could be made use of. The *Junzu zeli* includes a chapter prescribing how the accounts had to be balanced.[223] It was important that this was accompanied with documents giving evidence of the local situation, the layout of the logistics network and the exact distance of each logistics station from the next. It had minutely to be described when a military camp was relocated, in order to exactly calculate how much grain and equipment had been transported to the camps in what time frame. Such a list of the logistics stations is included in the precedents from the second Jinchuan campaign, yet the obligatory map of the logistics stations has unfortunately not survived.[224]

For each particular item in the ledger, be it money for the daily provisions, funds to buy transport animals or rice, it had clearly to be stated what amount of money constituted the primary funds or primary income, what amount of money was spent (*kaichu*), and what was left over (*shicun*) at the end of the accounting period. Each province kept ready a certain amount of money for ad-hoc military purposes that could at the beginning of a campaign be used to cover the initial cost. This amount of money was then accounted as the 'primary fund' (*chu'an*). Whatever was left over from the first accounting period was carried over to the next accounting period as 'old arrangements' (*jiuguan*). The income of the war chest was made up by transfers from the treasury of the war-waging province or by any transfers from other provinces or directly from that of the Ministry of Revenue. Remainders of grain were not allowed to be carried over to the next accounting period, so that these accounts were separate for each accounting period and not linked to the next. Lai Fushun says that this restriction was also valid for military equipment.[225] In case the Ministry did not accept an account it requested deducting the overspent money in the account and demanded that the responsible officials compensate the overspent sum, wherever they might take the money from. This process had to be recorded, meaning not only the original sum originally claimed from the accounting bureau was to be listed, but also that reduced by the Ministry, plus the amount of money to be paid back by the responsible officials. It was not permissible to simply replace the original account by a new account showing lower figures. Once the expenditure for all cost

---

223 *Hubu junxu zeli*, juan 9, fol. 10a–10b.
224 *Jinchuan junxu li'an*, juan 2, 91a–137a.
225 Lai, *Qianlong zhongyao zhanzheng*, p. 385.

groups was settled, the accounting bureau drew up the general account, submitted it, whereupon the ledger was finally closed.

During the Dzungar campaign of 1754 this method of accounting was not yet used, but expenditure was brought to account as soon as it accrued. It was the governor of Shaanxi, Hengwen, who managed the northern logistics route and suggested waiting until the end of the campaign before drawing up an account.[226] In the early years of the Qianlong reign there were still no clear rules for the accounting process nor did particular officially authorized prices for expenditure exist. After the end of the first Jinchuan war therefore, the Minister of Revenue Shuhede, his vice minister Zhaohui, and the governor-general of Sichuan, Celeng, deliberated about the accounting process and came to the conclusion that it would be best to use the precedents from the 1746 campaign against the native king of Djandui. Yet there was no comprehensive handbook about these precedents, so that a great deal of information was lacking. In many cases it was not possible to amend accounts that had been refused by the Ministry, and they were not again submitted, while for other cost groups no account had been drawn up or submitted at all. The provincial administration commissioner (*buzhengshi*) of the province of Sichuan had even been dismissed because of the confused state of the accounts. He was nevertheless reinstated and managed to create a kind of final account based on more than 2,000 bills.

The final accounting for the Dzungar campaigns proved quite difficult because the new campaign against the Khojas of Kašgar and Yarkant had begun before the account was closed. It was only in 1759 that the compilation of the final account for the six-year-long war in the west was resumed. The account of the western route was only closed in 1763. The general account included 498 items, a certain number of which had still not been accepted by the ministries of Revenue and of Works until it was finally closed in 1770. This means that the whole accounting process took twelve years, which was even longer than that of the second Jinchuan campaign.[227] The final accounting for the Myanmar campaigns was finished in a relatively short time. In 1771, only two years after the end of the last campaign, 180 items were merged into an officially complete account. This was only possible because experienced officials from the province of Gansu were employed that had already participated in the final account

[226] Mentioned in Lai, *Qianlong zhongyao zhanzheng*, p. 385.
[227] Lai, *Qianlong zhongyao zhanzheng*, p. 387.

of the western campaigns. They used the precedents from the war expenditure regulations from the Dzungar wars.

The campaign against Taiwan posed the problem of crossing the sea. This did not only affect communication between the fighting units and the commanders in Fujian, but also the organization of vessels to ship troops, provisions and equipment. It was especially the large amount of rice consumed during that campaign that led to a delay for the final account. This took more than two years to complete, though it only included 34 items. It had been produced when the empire-wide war expenditure code *Junxu zeli* had already been issued yet remained an exception because no rules concerning maritime transport had been provided for in the code. Another complex issue was the casting of money on the spot which served to supply the troops with sufficient cash to pay the sutlers with.[228] Coins had also been cast during the western campaigns, yet with the intention to replace the local *pul* coins (by the Chinese called 'red cash', *hongqian*, because of their high copper content) with Chinese ones.

As the campaign in Annam was not very extensive it posed no greater problems for the final accounting. For the two Gurkha campaigns, the issues to be brought to account were not very numerous, yet there was such a great distance between the war-waging province Sichuan and the war theatre that for the final accounting it had to be split into two parts, in order to simplify the calculation process.

The experience drawn from the western campaigns was a great help for the accounting of the second Jinchuan campaign. The court dispatched specially commissioned Grand Ministers who were personally to make sure that nothing went wrong. There were hundreds of accounts to be checked in detail for consistency and to make them tally with the respective regulations for war expenditure. After the second Jinchuan campaign, five accounts (*an*) were handed in as covering expenditure for 'barbarian affairs' at the beginning of the campaign, 399 accounts covered the period from the beginning of the war to August 1773 (the 'old files'), and 467 accounts (the 'new files') were set up for expenditure between that date and the 'triumphal return' of the troops in early 1776. To this total of 871 items a further 23 were to be added for disputed expenses.[229] 247 accounts concerned the payment of troops, 86 rewards and compensations, 24 the payment of staff in the logistics stations, craftsmen and family

228 Chinese coins were not minted but cast.
229 *Jinchuan junxu li'an*, juan 2, fol. 177b–178a.

allowances for labourers, 48 items horses and labourers in the courier stations, 233 items the transport of grain, 145 that of military equipment, and 89 the pay for civilian officials, the construction of roads and bridges and the building of boats, adding up to a total of 872 items (the heading says there were 877 items in total). The documents were counter-checked by the Ministry of Revenue and then either approved or sent back for resubmission, with the requirement to cut cost. The large number of files from the five logistics lines was chronologically divided into two parts. The first reason for this was that this facilitated the handling of the huge quantity of documents and the second that at the beginning of the war special rules, the so-called regulations for 'barbarian affairs', had been valid for war expenditure that were easier to cope with when separated from the main body of the files. The point of division was, nevertheless, the lost battle of Mugom in the summer of 1773, after which larger contingents of troops, especially Banner troops, were dispatched to the front. To them different rules of payment from those for the troops applied who had been brought into the war theatre until that date. The participation of troops from many different provinces made it all the more necessary to create an overall valid war expenditure code for future campaigns. The accounting bureau in Chengdu was split up according to the five logistics lines. Each division was headed by a circuit intendant (*sidao*). The files of each logistics line were again split up into stations inside (*kounei*) and outside provincial territory (*kouwai*). It might be well to remember that in the late 18th century the western border of the province of Sichuan was just on the eastern slopes of the Qionglai Range, where Eastern Tibet began. For the accounting process, precedents from both the first Jinchuan war and the western campaigns were used, in some cases also precedents from the Myanmar campaigns.

The whole auditing process for the second Jinchuan war lasted almost a decade, and the account was only closed in 1784. It was then realized that not much of the huge amount of money allocated for the war, was left.[230] The long process of checking, counter-checking and revising the bills had actually been prepared quite well and had started rather early. All items necessary to feed, pay and equip the army were organized by the logistics bureau which for each item or service purchased had to draw up an account to be submitted to the Ministry of Revenue for approval. The bureau could, of course, not delay the payment until they had received

230 *Jinchuan junxu li'an*, juan 2, fol. 187a.

the Ministry's answer, as the immediate demand of thousands of troops and civilian labourers had to be met. The consequence was that there was always an imminent conflict between immediate expenditure and later approval or obligation to reduce sums in the account.

It could be seen in the foregoing sub-chapters that the regulations for many items widely differed from province to province and from war to war. In most of the large campaigns of the eighteenth century troops from many provinces were involved and paid differently from those of other provinces. The task to ascertain which rules were valid in which cases, was not easy. In March 1775 therefore, the emperor dispatched several high officials, Ebao (director-general of grain transport), Guilin (provincial commander of Sichuan), vice minister Liu Bingtian, Hao Shuo (governor of Shandong), and Yan Xishen (governor of Hunan), to support the Minister of Revenue Fulehun and Wenshou (governor-general of Sichuan) with the frightening task to sort out the accounting and to disclose cases of embezzlement.[231] Li Hu, provincial administration commissioner of Yunnan, was later specially entrusted with the revision of accounts that had been sent back by the Ministry of Revenue. Responsibilities were shared among the six persons. Ebao, for instance, was appointed to manage the cost of the grain transported along the northern route. A year later, just when the war was ended, only 40 accounts had been submitted and were approved by the Ministry.[232] The other accounts drawn up until then were constantly sent to and fro, and there seemed to have been an atmosphere of permanent negotiation between the Ministry and the war expenditure bureau in Chengdu about trifling matters.

The reasons for the long period of confusion are outlined in a memorial quoted at the end of the *Jinchuan junxu li'an*.[233] Points of discussion were whether old or newly established rules were to be adhered to, whether the regulations for nearby or distant stations were in force, whether the overall conditions had changed or not, or whether the rules for inside provincial territory rather than those for outside provincial territory were to be applied. Besides these questions, another field of trouble was that of the several hundred civilian officials working in logistics and therefore entrusted with accounting matters. Quite a number of them had left their

231 *Jinchuan dang* 40/I/00119 (QL 40/2/9).

232 *Jinchuan dang* 41/I/00035 (QL 41/1/6).

233 *Jinchuan junxu li'an*, juan 2, fol. 190a. There is neither an author nor a date cited. The memorial must have been compiled in the early 1780s.

post in the province of Sichuan or had died or retired in the meantime, so that it was impossible or at least very difficult to have them interviewed.

One example might highlight the difficulty in which the logistics managers found themselves during the war and when the time came to submit their ledgers to the Ministry of Revenue: A sum of about 93,000 *liang* was brought to account for tents and cooking woks for imperial soldiers, auxiliary soldiers and labourers. The argument of the officials claiming this sum was that the labour cost for making tents and other kinds of living equipment had doubled since the years of the first Jinchuan campaign, so that the prices of the precedence cases from that period could not any longer serve as a basis more than twenty years later. Yet the accounting rules did not allow to bring tents to account which were made for native auxiliaries and for hired personnel, and not for regular troops. On the other hand, the auxiliary troops and the other personnel had to follow the imperial troops to the frontline, where the weather conditions were really harsh during the winter, and it would hardly be possible to let them dwell outside in wind and snow. Therefore the Ministry had decided that the cost for producing tents for those kinds of troops should be deducted from their pay. Yet the problem was that the responsible native king had never sent their pay so that there was nothing to be deducted from. The servicemen meanwhile had to follow the newly advancing frontline, where so far no logistics stations had been established after the quickly advancing army had passed. Because the logistics staff had to liaise with the army as soon as possible, there was not even time for them to erect sheds or other accommodation, which meant that they had to make do with tents. A third reason why the responsible officers had stated the tents in their account was that during the battle of Mugom many tents had been lost to the enemy. The Ministry nevertheless accepted those reasons only in part and rejected the account. As the responsible officials had already requested further cooking woks and tents after the Mugom catastrophe, the Ministry accepted that the case was urgent and therefore allowed the item to be appended to the general account as a matter of compensation concerning money which had to be paid back by the responsible civilian officials duty.[234]

There is a long appended account (*fuxiao*) in which items are listed that had been paid for unduly.[235] An amount of 2,620,596 *liang* and the

---

234 *Jinchuan junxu li'an*, juan 2, fol. 148b–149b.
235 *Jinchuan junxu li'an*, juan 2, 142a–171b.

cost for 74,382 *dan* (7,699 m^3) of grain were to be paid back by officials that were held responsible for overspending. The reimbursed money was called 'subsequent income' (*xushou*). The officials had balanced their expenses with 415,712 *liang*, and the emperor magnanimously renounced the missing 2,415,567 *liang*. Some of the officials had died in the meantime, and some were simply insolvent. The emperor also waived another 986,954 *liang*, which could not be paid back by merchants whose porters had run away with the grain. In those cases either the entrepreneurs or their families were bankrupt and unable to replace the loss. Nevertheless the emperor ordered to pursue them as far as possible and to try to retrieve the lost money wherever possible. A third group of persons owing money to the state were civilians unable to pay the *jintie* surcharge through which the transport expenditure for the tax silver or tax grain was normally financed, which was in fact nothing else than a special kind of tax levied on the field-and-poll tax. The peasants had therefore borrowed money from the provincial treasury which was to be paid back in instalments over a period of three years, beginning with the end of the war. When the final account was rendered, 1,543,814 *liang* were still missing, and the emperor renounced the payment of this sum.[236] The emperor also ordered to waive some money missing due to converting quantities of rice from charity granaries (*yigu*) into the equivalent monetary value. Part of the money had already been paid back and received when the final account for the war was compiled, but 13,789 *liang* were still missing. The emperor then decided to relinquish that money. A similar case was the lead borrowed from the provincial mint and other state-owned workshops to the value of 23,963 *liang*. In this case no repayment was requested. Other provinces had provided the money for baggage pay and the purchase of horses. For those items, Sichuan province—responsible for the military expenditure in Jinchuan—brought to account expenditure amounting to 291,379 *liang*. This sum had to be paid back to the provinces which had provided the respective sums. Sichuan province itself paid 162,922 *liang* as baggage pay to the troops marching to Jinchuan. Furthermore, loans to civilian and military officials from all types of troops and to merchants working in the logistics lines amounted to 150,935 *liang*. These loans, as well as the baggage pay, were not to be paid back, while other loans amounting to 58,831 *liang* and granted for baggage pay, the purchase of horses and the return transport of wounded soldiers as well as corpses of

---

236 *Jinchuan junxu li'an*, juan 2, fol. 178b.

Banner soldiers, were to be paid back in instalments over 3 to 10 years. This procedure is in contradiction to the later rule that the Banner troops did not have to pay back their baggage pay. A last item waived was a sum of 33,339 *liang*, left over from the purchase of additional husked grain from some granaries with a value of 680,070 *liang*.[237]

In total the emperor had waived the repayment of 5,294,028 *liang*. If he had not been so generous—forced by the circumstances—this large amount of money would actually have been paid by civilian officials working in the logistics lines. This was the situation when the account of the expenditure for the second Jinchuan war was officially closed in 1784. Yet the discussion was not finished yet. There were some 358,857 *liang* still missing in the provincial treasury, owed to the province by various civilian officials. On top of this amount, a sum of 37,041 (or 34,822) *liang* was due to cover expenses for iron that had not been retrieved from the camps and the battlefield.[238] The regulation to leave the iron where it was seems to be a decision that was made during the auditing process for the second Jinchuan war.[239] From spring 1781 on the provincial government had begun to collect debts by deducting money from the *yanglian* pay of officials that had worked in the field of logistics and could be made liable. This was not an easy task because many of them had in the meantime been appointed to posts in other provinces. After all, 121,492 (123,709) *liang* could be recovered by this means until 1784, through a method that was so promising that an official submitting a memorial calculated that after nine years all missing money could be recovered in this way.[240] While the emperor often showed clemency to 'small fry', he was adamant towards officials that had climbed the ladder of official career or had become rich because of their successful involvement in the organisation of the campaign,[241] and promised to issue a special edict for the retrieval of funds from officials working in other provinces. Debts produced by state officials were, the emperor said, a very different affair from money that was owed to the government by private persons or even merchants. There were apparently specialized statutes that regulated the repayment of money to the state as a duty distributed on the shoulders of several persons (*fenpei*

[237] *Jinchuan junxu li'an*, juan 2, fol. 178b–179b.
[238] *Jinchuan junxu li'an*, juan 2, fol. 189b, 196a.
[239] *Jinchuan junxu li'an*, juan 2, fol. 44b.
[240] *Neige daku dang'an* 090688 (QL 50/5/16).
[241] Dai, "Military finance", pp. 309, 311.

zhi li).[242] The money thus collected from accountable officials was to be regarded as a contribution (*juan*) to the war expenditure. A very typical case is that of a certain Wang Ting, a sub-prefectural magistrate (*tongpan*) somewhere in the province of Anhui who had participated in the war as an accountant and was made liable for 1,363 *liang* but who was already dead in 1788.[243] An exact investigation that was carried out to discover private property or hidden money, came to the result that there was only his destitute widow and a ten-year-old son, with an elder brother living far away. Thus the proceedings could not but be closed. The emperor tended to spare lower-ranking officials like local government assistants (*zuo'er, zuoza*) that were less prosperous. He furthermore argued that the price of iron was so low that it was not worth taking the trouble to recover the 30,000–odd *liang* missing. Overzealous accounting managers in Sichuan had even deducted money from the salaries of officials that had not taken part in the campaign and were only dispatched to Sichuan later on, when the final account was to be compiled. Promises were made to pay back this sum, amounting to 67,271 *liang*.[244] In order to be able to close the final account, the emperor was asked to renounce the repayment of the remaining 257,522 *liang*.

The discussion about back-payments in Sichuan went on. It was mostly a question of smaller amounts, as could be seen above, that continued occupying those that had participated in the war. In 1790 the governor-general of Hu-Guang claimed assistance from the province of Sichuan for invalid soldiers that had been injured during the war and were hardly able to make a living. This case still occupied the authorities in the early years of the Jiaqing reign.[245] In the same year garrisons in the province of Gansu claimed back money from Sichuan for riding horses and mules for 'corporals' (*waiwei*) and common troops that had been purchased at the end of the war on returning home.[246] A case similar to that of Wang Ting

---

242 *Jinchuan junxu li'an*, juan 2, fol. 194a. The rules for collecting the individual share of officials are mentioned in Shen Shucheng's overview over the precedent rules of all government institutions. The *fenpei* rule was particularly applied for officials of the local administration that were personally made liable for missing amounts of money. Shen Shucheng, *Zeli bianlan* (print from 1791; electronic version [data bank] used via licence by Staatsbibliothek Berlin of Zhongguo jiben guji ku, ed. by Beijing Airusheng shuzihua jishu yanjiu zhongxin, Beijing), juan 25, paragraph *Cangni kuyin jianie kuikong* 'Feigning the lack (of funds and grain) in [district or prefectural] granaries and treasuries'.

243 *Neige daku dang'an* 067698 (QL 53/10/27).

244 *Jinchuan junxu li'an*, juan 2, fol. 193b.

245 *Neige daku dang'an* 068885 (QL 55/6/18), 062007 (JQ 6/12/19).

246 *Neige daku dang'an* 049156 (QL 56/11/18).

turned up in 1793 in Zhejiang, when a certain Chen Yushang, formerly district magistrate (*zhixian*) of Changshou in Sichuan, was made liable for 2,671 *liang*.[247] The loan for the baggage pay of the troops from Hubei and Sichuan was also not completely paid back by the soldiers years later.[248] It can thus be concluded that although the final account for the expenditure of the second Jinchuan campaign was officially closed in 1784, a lot of liabilities there were still not settled either between various government agencies (province–province, province–provincial mint) or between officials or soldiers and the government.

Conclusions

The whole accounting process was, as could be seen throughout this chapter, subject to strict bureaucratic regulation, or, as Dai Yingcong has said, "bureaucracy was in command".[249] This was basically not a bad method because a system of official prices for objects and services provided both the managers of the war logistics bureau and the Ministry of Revenue, who in the end footed the bill, with a reliable basis upon which expenditure could be settled without a complex discussion. The intricacy of the accounting was rather due to the fact that there was, at least until the late 1770s, no empire-wide price regulation for war expenditure. Instead, each province had its special regulations to which the troops from that particular province were subject, instead of to the war expenditure regulations of the province that was responsible for the campaign. Furthermore, local conditions of each campaign made it impossible to use the regulations of earlier campaigns as precedents in all respects. This was particularly true for the prices of grain, beasts of burden, the cost of transport, as well as individual rewards. For grain and animals, there were, at least, rules of thumb established in earlier precedents, but they were never transformed into law as always being subject to local market conditions. The cost for transport, on the other hand, could be and was regulated by the *Junxu zeli*. This was much necessary because transport cost during the second Jinchuan war had skyrocketed to an unexpected height—admittedly also because of the exceptional topographic conditions. Transport cost of food and military equipment had, as could be seen, made up nearly

---

247 *Neige daku dang'an* 063309 (QL 59/2/11).
248 *Neige daku dang'an* 717432 (QL 54/6), 002560 (JQ 1/10/4).
249 Dai, "Qing State, Merchants", p. 37.

80 per cent of the total expenditure for this war. This might admittedly have been lower during other campaigns that were fought under more favourable topographical conditions. Reducing expenditure for war could only be made possible by paying greater attention to a standardization of official prices for goods and services, leading to the compilation of the War Expenditure Code *Junxu zeli* in 1785, and to a better organization of the logistics lines. Before this was made possible, the usual means of raising funds were to be applied to cover the excessive cost for war. For this purpose the Qing had introduced the formerly unknown war supplies and expenditure bureau (*junxuju*) that on the one hand took care of transport, food and equipment and on the other hand strictly documented and controlled income and expenditure of the war chest.

The philosophy of the early Qing emperors was to "nourish a strong military".[250] Cost could therefore neither be cut in the salaries nor by reducing the size of garrisons. The "imperial favouritism" of the Kangxi emperor even went so far as to introduce a stipend for soldiers with which to engage in entrepreneurial operations.[251] This was only possible because the state revenue had been consolidated and the first problems of a war-induced 'national' deficit that occurred during the Shunzhi reign changed into a huge surplus during the Qianlong reign with which wars could be financed without resorting to extraordinary means of fund raising. This situation changed again during the White Lotus rebellion, when the military leaders as de-facto war time profiteers virtually plundered the state treasury.[252]

---

250 Dai, "Nourish a Strong Military".

251 Dai, "Yingyun shengxi".

252 Dai Yingcong, "Civilians Go into Battle: Hired Militias in the White Lotus War, 1796–1805", in *Asia Maior*, Third Series, 22/II (2009), p. 177.

CHAPTER TWO

WAR FINANCE

> The contribution campaign is of little help for military expenditure and only an occasional means of fundraising. My Father [the Kangxi Emperor] had in the past repeatedly stressed that contributions were no praiseworthy matter.[1]

The organization of war expenditure was settled by the 'war-waging' province that for this purpose set up a war chest separated from the normal budget, yet fed with funds from the 'civilian treasury', so that direct transfers from the civilian treasury to the war chest were to be deducted from the provincial revenue retained in the province. Loans by other provinces had to be balanced in the framework of inter-provincial cash flow, for instance, with funds for disaster relief or the purchase of grain to stock up the district granaries. Advances from the Ministry of Revenue to the war chest were charged against the annually due tax remittances. Yet in the end the greatest part of the cost was nominally taken over by the Ministry of Revenue in the central government, which means that the war-waging province and the neighbouring provinces supporting the former were granted tax-abatements.

Further on it will be seen that these meticulously documented funds only constituted part of the money by which war was financed. Most important is that the Qing government, unlike its predecessors, the Ming, abstained from levying extraordinary taxes to fill the war chest, except for a few cases during the early decades of the dynasty. At the same time it operated according to the axiom that military expenditure had to be covered by the government and out of government revenues, and not with the help of credit. The Qing state was in the eighteenth century financially capable not only to do without government debt, but also to lend money to the troops to finance their equipment, to merchants that were required to contribute financially to the war as a kind of tax surrogate, to state officials that were made liable for overspending, and to the common people in the shape of tax extensions, cuts or abatements. It does not seem that

---

1 *Pingding Zhungar fanglüe, Qianbian*, juan 14, fol. 27a (YZ 2/9/guimao).

the Qing state around 1800 was approaching bankruptcy in any case, in spite of one and a half century of permanent warfare and irrespective of two thirds of the state revenue being spent for the military. Even in the provinces the depletion of treasuries in the late eighteenth century was not attributed to military spending but to mismanagement and corruption, like in the case of Fukang'an who, during his reign as governor of Sichuan and then of Guangdong-Guangxi misappropriated government funds to such a degree that he is often compared with the emperor's notoriously corrupt minion Heshen.[2] Similarly famous is the huge misappropriation of contribution funds for disaster relief in the province of Gansu that was disclosed in 1781.[3]

In this chapter it will be explained how the fund-raising to pay the military campaigns of the Qing state worked in practice, and what parts of the war expenditure were—in spite of all rhetoric—not taken over by the government but were so regularly shifted on to the shoulders of private persons, mainly state officials, that a system of involuntary common share had come into being until the early nineteenth century. While the provincial governments took over the largest part of the burden, supported by direct remittances from the Ministry of Revenue, temporarily limited contribution campaigns yielded a substantial amount of money that came from the private purses of state officials or merchant associations. The income from these institutionalized contributions was seen as governmental revenue, for which reason the amounts of money collected through these campaigns are in most cases known, at least those of the larger payments. Yet the share to war expenditure covered by government officials that had to vouch for overspending is as unknown as the real cost of each of the seventeenth and eighteenth century wars, because the government was only willing to pay what had been brought to account according to the regulations explained in the foregoing chapter, and not one penny more.

The first part of this chapter will provide an overview of the instruments of war financing in the early and high Qing period; the second part will focus on the peculiarly Chinese system of institutionalized contributions, and the third part will offer a detailed insight into the fully developed practice of financing war during the second Jinchuan campaign (1771–1776), the most expensive war during the Qianlong reign.

---

[2] Zhaolian, *Xiaoting zalu* (Beijing: Zhonghua shuju, 1980), juan 4, p. 105.
[3] Guy, *Governors and their Provinces*, p. 219.

War Finance in the Seventeenth and Eighteenth Centuries

The most critical issue of war finance is that military campaigns are enterprises whose cost can hardly be calculated in advance. The financial managers of the state and its army can at least try regulating cost that accrues in any case, so that the multiplicands are known, even if the multipliers are the result of a variety of factors. In imperial China the largest part of expenditure was subject to strict regulations, as could be seen in the first chapter of this book. These regulations varied from province to province and from war to war up to the late 1770s. Rules for expenditure were written down in a campaign and then later used as precedents for military campaigns in the same region, but also for wars that were waged under different topographical conditions. There were, for example, precedents from the western campaigns, precedents from Yunnan (the campaigns in Myanmar) as well as those from the campaigns to put down the unruly mountain tribes in the west of Sichuan, two of which led to the first and the second Jinchuan war. These precedents for their part evolved from the earlier wars in eastern and central Tibet and in Qinghai. The War Supplies and Expenditure Code issued in 1785 nevertheless did not cover all types of expenditure, and some are missing even among the minutely listed ranks of Banner officers and civilian officials of the central government that might take part in a campaign.

It is especially the regulations for objects like military equipment, horses and rice that were more open for market conditions that differed from province to province. Although the Qing administrators were often indulging in "overcentralisation by the prospects of greater administrative rationality" or in the lore that "information is control",[4] practical Qing bureaucrats were well aware of the limits of state power and even knew that merchants were not purely to be regarded as selfish hoarders and profiteers. The price mechanisms of the market were by some observers recognized as often more effective than state planning, as Helen Dunstan has shown. Local grain vendors, for instance, stored grain and threw it on the market in times of deficient harvests, supplying their community with the grain they were in dire need of, while the untimely hoarding activities of the state often only exacerbated the price increases of grain.[5] The local

---

4 Helen Dunstan, *State or Merchant: Political Economy and Political Process in 1740s China* (Cambridge, Mass.: Harvard University Press), p. 3.

5 Dunstan, *State or Merchant*, pp. 464–466.

and provincial administrations furthermore derived much of their operating revenue from locally levied taxes on commerce,[6] and were therefore not interested in curtailing mercantile activities on which the vital functions of the state and its executive instruments depended, especially in times of hardship like drought, inundation, or war.

The aim of all the regulations for war expenditure was to cut cost as far as possible. This can be considered as a general hint that war should be, in the conception of the Chinese state, as cheap as possible, at least as seen from the viewpoint of the Ministry of Revenue in the central government. But this does not mean that the absolutist states of Europe were not in favour of lowering the cost of military campaigns.[7] As in China warfare was part and parcel of the everyday bureaucracy it was as such also subjected to the financial customs and procedures in normal administration. In this respect the late imperial Chinese state was quite successful in adopting the method of the "bureaucratization of violence" as described by William McNeal.[8] This is also true for war finance. Unlike in Europe, there was no budget of 'affaires extraordinaires',[9] like in absolutist France, not to mention a concept of government bonds as developed in Great Britain with the foundation of the Bank of England.[10]

Taxes and Various Levies and Duties

But where did the money for the numerous military campaigns of the early and high Qing periods come from? For the suppression of the Three Feudatories in the late 17th century a sum of 1.87 million *liang* was derived from the field tax of the provinces of Zhili (modern Hebei), Henan, Jiangxi, Anhui and Zhejiang, 300,000 *liang* were used from the income of the salt tax in Zhejiang and Liang-Huai (Anhui and Jiangsu north of the Yangtse River), 400.000 *liang* came directly from the treasury of the Ministry of Revenue, and the rest were 'miscellaneous items' (*zaxiang*) from Henan,

---

[6] Guy, *Governors and their Provinces*, p. 301.

[7] Van Creveld, *Supplying War*, p. 39.

[8] McNeal, *Pursuit of Power*, p. 117.

[9] Concerning war finance in eighteenth-century France, see for instance, James C. Riley, *The Seven Years War and the Old Regime in France: The Economic and Financial Toll* (Princeton: Princeton University Press, 1986).

[10] The relation between the foundation of the Bank of England and war finance is demonstrated in the introductory part of the Bank of England Act from 1694: "Wee [...] assembled in Parliament for the further Supply of your Majesties extraordinary Occasions for and towards the necessary Defence of your Realmes [...]". See http://www.legislation.gov.uk/aep/WillandMar/5-6/20/introduction.

Shandong, Jiangxi, Zhejiang and Jiangsu.[11] This amounted, of course, only to a tenth of the whole expenditure of about 30 million *liang*, but the sources were similar to those of later wars. In this example, about 10 per cent of the cost were covered by the treasury of the Ministry of Revenue (*buku*) directly, 8 per cent came from the salt tax (*yanke*), and 47 per cent were met by the provincial treasuries (*fanku, siku*). 'Miscellaneous items' constituted 34 per cent of the total cost. It is unfortunately not said what exactly the 'miscellaneous items' were, but it might be assumed that they were either various surcharges or probably contributions (*juan*) from the private purses of provincial officials.

Concerning the so-called ten successful wars of the Qianlong emperor, Lai Fushun has compiled a table that includes information about the origin of the funds needed for these wars.[12] The first Jinchuan war, costing 9.964 million *liang* according to a document in the archives of the State Council (*junjichu*), was financed by the province of Sichuan to the tune of 772,900 *liang*, and the rest of 8.791 million *liang* was paid by the Ministry and the provinces of Hubei, Hunan, Henan, Jiangxi, Shanxi and Guangxi. The burden of 35 million *liang* spent for the western campaigns was likewise distributed on the shoulders of various provinces and those of the Ministry, yet more exact data are lacking. The Myanmar campaigns, costing 13.2 million *liang*, were financed with the help of eight transfers, the Ministry supplying 3 million, while the rest came from the provinces of Henan, Anhui, Jiangsu, Shandong and Zhejiang, as well as the salt administration zones of Changlu (near Tianjin) and Liang-Huai (northern Jiangsu), and from contributions by private persons. The Taiwan campaign, which cost 8.9 million *liang*, was paid with the help of transfers from the Ministry and provincial treasuries. 29,780 strings of coins were transported to the island, as well as 4.2 million *liang* of silver.[13] On the island itself, an additional 90,000 strings of cash were produced to pay for the food for the troops. The Annam campaign is of particular interest because a small sum of 2,800 *liang* was 'collected' in Annam, as well as 2,000 strings of cash, presumably as contributions from the local gentry. The province of Guangxi delivered 1.06 million *liang*, and Yunnan 300,000 *liang* to cover the whole expenditure of 1.3 million *liang*. Of the 1.5 million *liang* needed for the first Gurkha war, Shanxi and Shaanxi provided 1 million, and

---

[11] Quoted in Chen, *Qingdai junfei yanjiu*, p. 246.

[12] Lai, *Qianlong zhongyao zhanzheng*, pp. 426–428.

[13] It was custom in traditional China to transport huge numbers of coins by threading up 1,000 cash on a string (*guan*). The 'string' thus also served as a unit of account.

Hunan the rest. The second Gurkha war, costing 1.75 million *liang*, was financed by Sichuan with 1.05 million, and 6.7 million *liang* came from other provinces. The discrepancy in the last case between the total cost and the funds raised is not sufficiently solved. Apparently a much larger sum had been allocated for the war than was really needed. It might be guessed that the experience during the second Jinchuan war with the problem of the huge cost for the transport of rice and equipment in high mountain territory had caused the logistics managers of the second Gurkha war to anticipate high cost rather than to risk facing problems with the supply. As this money, furthermore, came from contributions it was no direct burden on the state budget.[14]

There was also another means of war finance better known in the west, especially in Prussia, where the attempt was made to gain higher revenue by issuing debased coins, which allowed the government to obtain a higher margin of profit from the seigniorage.[15] In Qing China, increased production of coins also played a certain role, but was not decisive for war financing. Chen Feng, who has analysed the data of the provincial mints in 1651, came to a sum of 113,518 *liang* yielded annually by the seigniorage (*zhuxi* 'casting profit'). It was intended to use this money directly as pay for the troops.[16] Richard von Glahn provides an overview on the seigniorage revenue of the first 15 years of the Qing period and shows that its average annual revenue decreased from about 372,000 *liang* in the first five years to 210,000 *liang* in 1651 and finally to 68,000 *liang* in 1654.[17] This means that in 1651 about half of the income from the seigniorage was, according to Chen Feng, spent for war expenditure. Yet generally speaking, the imperial mints throughout the empire were not able to operate profitably. After 1700 the mints even produced substantial deficits that could by no means be met by decreasing the copper content of the coins, and even after the relatively cheap copper from Yunnan was available

[14] Research on contributions in the 19th century has been done by Elisabeth Kaske, "Price of an Office", and "Fund-Raising Wars". The most important Chinese study on the contribution system during the Qing period is Xu Daling, *Qingdai juanna zhidu* (Xianggang: Longmen shudian, 1968).

[15] Compare the article by A. O. von Loehr, "Die Finanzierung des Siebenjährigen Krieges: Ein Versuch vergleichender Geldgeschichte", in *Numismatische Zeitschrift* XVIII (1925), pp. 95–110.

[16] Chen, *Qingdai junfei yanjiu*, pp. 335–336. Copper cash in China was not minted but cast.

[17] Richard von Glahn, *Fountain of Fortune: Money and Monetary Policy in China, 1000–1700* (Berkeley: University of California Press, 1996), p. 210.

from the 1730s profits did not exceed five per cent and amounted to less than a thousandth of the total tax revenue.[18]

Another, often overlooked means of raising funds was a surcharge levied upon the transport of grain or other items. In the early Shunzhi reign it was suggested that part of the so-called 'tribute rice' (*caoliang, bailiang*) transported from the granaries in the lower Yangtse region along the Grand Canal to Beijing, was partially converted into money, according to an exchange rate that ranged between 1.4 and 2 *liang* per *dan* of rice. This means that the tributes of these prefectures were half paid in rice, and half in money, which was branched off to pay the army. The taxes to be delivered in this way were paid in full, while the taxpayers were saved half of the transport cost of the grain. This was at least the convincing argument of the Ministry of Revenue for the application of this method. In this way, 352,000 *liang* could be counted on as income of the state treasury, and in the following years the same method was used again to yield 3.6 million *liang* per year.[19]

A similar, later frequently used, method was also derived from the transport system, namely the surcharge on the difference between the lighter provincial weights (*yuping*) and the heavier capital weight (*kuping*).[20] For this reason, on all items purchased for the army a surcharge of one per cent (called *yuping* or *pingyu*) was levied, except for the funds allotted for salaries and daily provisions (*yancai*).[21] The considerable income from this surcharge, namely between a half and one per cent of the commodity value according to some regulations, was to be used to cover excessive expenditure.[22] It is worth mentioning that this surcharge was in the end not only levied to cover the difference between the lighter weights in the provinces and the heavier weight units in the capital with which the Ministry calculated, but also in order to raise additional funds. During the Jinchuan campaign the so-called 'lubrication subsidy' (*jintie*) or 'assistance subsidy' (*bangtie*) was introduced through which the transport of grain was partially financed. It was from then on used regularly, not only

18 Hans Ulrich Vogel, "Chinese Central Monetary Policy, 1644–1800", in *Late Imperial China* 8/2 (1987), pp. 11, 15.

19 Chen, *Qingdai junfei yanjiu*, pp. 336–337.

20 The *kuping* weight in fact differed from province to province. In Guangdong, for instance, the *kuping* was 0.4736 g heavier than the capital *kuping*, while that in the prefecture of Ningbo was 0.4463 g lighter. Huang Da, Liu Hongru and Zhang Xiao (ed.), *Zhongguo jinrong baike quanshu* (Beijing: Jingji guanli chubanshe, 1990), vol. 1, p. 93.

21 *Hubu junxu zeli*, juan 9, fol. 9b–10a.

22 Lai, *Qianlong zhongyao zhanzheng*, p. 391.

in the Gurkha campaigns, but also for the suppression of the White Lotus rebellion during the Jiaqing reign.

Irregular Levies

The methods described above appear quite sophisticated in contrast to some others used during the Shunzhi and the early Kangxi reign to finance the military campaigns, namely advance levies of the field tax, extraordinary taxes, raising taxes, intensified reclamation of wasteland and detailed examination of land registers, as well as a reduction of expenditure by cutting salaries, allowances or commissions. To these methods, irregular contributions have to be added by which individual officials were compelled to deliver funds, rice or to donate transport animals to the army. The difference to the institutionalized contributions was that the donators in this case could not expect to be granted something in return (for instance, a brevet title or an option for a higher office), as it was customary for the regular *juanna* and *juanshu* contributions, the functioning of which will be described below.

Enforced contributions (somewhat similar to the Thirty Years' War *kontributionen*) were used especially during the Shunzhi reign, when the army took grain, fodder and transport animals from the local population as needed. The arbitrary requisitioning of food, fodder, boats and carts was soon recognized as a factor for conflicts between the population and the government, represented by the ethnically different Manchus. The Ministry of Revenue therefore required that the expenditure for these items were shouldered by state officials. It was also ordered that the cost for food and services might properly be brought to account and be balanced by government institutions. The same was valid for sickles, horse troughs, buckets, woks, nails, cooking oil, coal, or firewood that was freely commandeered wherever the army marched through, and if paid at all, only at a ridiculous price. On top of requisitioning food, tools or animals, the armies marching through the provincial districts also required labour from the local population, in order to transport the baggage, prepare the camps or care for food and other items that might be purchased by the troops.[23] Chen Feng says that under the Kangxi emperor the situation improved somewhat, but it can be demonstrated that during the Qianlong reign it was still necessary to exhort the troops not to underpay the sutlers

[23] Chen, *Qingdai junfei yanjiu*, pp. 289–293.

in the camps—but also to admonish the sutlers not to fleece the soldiers that had no chance to buy rice, flour, tea, water, hay, beans, firewood or charcoal elsewhere.[24] It was also common in the late eighteenth century to grant tax abatements to the districts the troops had crossed on their way to the war theatre. A reasonable explanation for this cutting of tax might be that even at that time the columns of thousands of marching troops were detrimental to the local economy.

While tax increases played an important part in war financing especially in contemporary Great Britain,[25] the Chinese emperors of the eighteenth century wholly abstained from this means of financing their wars. Their political rhetoric was not to overburden the people and all emperors therefore refrained from openly increasing the field tax. The government therefore relied on other means of increasing the revenue from various kinds of taxes. Advance payments had already been common in the late Ming period, at a rate of a fifth or a third of the tax quota of the next financial period. The tax was to be delivered either in kind (rice) or in cash, yet both methods could increase the grain price up to four hundred per cent. In the early Shunzhi reign there were several instances that the whole quota of the coming year was to be paid in order to feed the army or provide funds to finance war. In Hunan, for instance, the pay, food and fodder for several garrisons whose troops were in the field was organized by requiring an advance payment of 31,000 *liang*, in Hunan by advance payments to the tune of 22,000 *liang*. During the Kangxi emperor's war against the Three Feudatories, Hunan advanced taxes amounting to 600,000 *liang*, and Shandong and Anhui 200,000 *liang* each.[26] The pay of troops campaigning was also financed with funds originating in the autumn harvest, as well as from the brokerage tax (*yahang shui*), the real estate transfer tax (*tianfang qi shui*), the mortgage tax (*dianshui*) or pawnshop tax (*dangshui*), commercial taxes (*shangshui*), and from various levies on cattle and other agricultural products,[27] alcohol and tobacco.

---

24 *Hubu junxu zeli*, juan 9, fol. 1b–3a.

25 John Brewer, *The Sinews of Power: War, Money and the English State, 1688–1783* (London: Hyman, 1989), pp. 30, 96–99. Larry Neal, "Interpreting Power and Profit in Economic History: A Case Study of the Seven Years War", in *Journal of Economic History* XXXVII/1 (1977), p. 32. D. W. Jones, *War and Economy in the Age of William III and Marlborough* (Oxford: Blackwell, 1988), p. 317.

26 Figures according to Chen, *Qingdai junfei yanjiu*, p. 295.

27 The pawnshop tax was normally 5 *liang* per shop (annually?), but in 1674 twice that amount was levied. See Huang Liuhong, *Fuhui quanshu* (print by Jinling lianxi shuwu, 1696; electronic version [data bank] used via licence by Staatsbibliothek Berlin of Zhongguo jiben guji ku, ed. by Beijing Airusheng shuzihua jishu yanjiu zhongxin, Beijing),

The Qing also inherited various extraordinary taxes introduced during the late Ming period and known as funds for the salaries in the Liaodong campaign (*liaoxiang*), for the 'bandit suppression' (*jiaoxiang*), and for training troops (*lianxiang*). The Liaodong tax was in some provinces levied according to land size, and for each *mu* (614 m^2) of land nine cash (*li*) or 0.009 *liang* were to be paid, in others the tax was levied according to the harvest volume.[28] This wartime surcharge was levied in all provinces except in the capital and in Guizhou. In the two provinces of Hubei and Hunan, for instance, the normal field tax (*diding*) was as high as 467,000 *liang* in 1657, while the 'nine-cash-tax' produced a revenue of 273,000 *liang*. In 1661 the extra war tax collected in seven provinces yielded more than 5 million *liang*. The province of Zhili alone paid 451,000 *liang* in taxes as 'salary for training troops', and an amount of 27,000 *liang* as 'irregular salary for training troops' (*ewai lianxiang*).[29] Chen Feng has demonstrated that these extra levies provided seventy per cent of the pay for campaigning troops. Other sums collected as taxes were directly allotted for the payment of horses, tools and transport. In the province of Jiangsu it was very common to impose special taxes on fields owned by state officials, persons who had passed the highest level of the state examinations or were enrolled in the National University (*taixue*) that was supervised by the Directorate of Education (*guozijian*). Persons belonging to these groups were normally not taxed, yet the Kangxi emperor was in such dire need for funds to finance the war against Wu Sangui that he allowed the collection of a field tax for state officials (*guanhuan diding*) in several districts, at a rate of 0.03 *liang* per *mu* (approx. 614 m^2) of land and 3 *dou* per *dan* of rice harvested (30 per cent), which was twice the amount of taxes to be paid by peasants. For the western campaigns during the Kangxi and Yongzheng reigns the provinces of Gansu and Shaanxi, that were strained anyway with the organization of the campaign, were ordered to deliver an additional field tax of 3 cash per 0.01 *liang* of tax, and three per cent more of grain. The province of Sichuan also levied additional taxes with an amount of 610,000 *liang* to support the current campaign.[30]

juan 8, paragraph *Zake bu, Dangshui* [Part miscellaneous taxes: Pawnshop tax]. The livestock taxes (*luodi shui*) were levied on the sale of cattle in another district.

[28] Dai, "Yingyun shengxi", p. 6, footnote 19, mentions a case from 1661 when the Shunzhi emperor planned to levy a 'militia surcharge' of 1 *fen* per *mu* of land, but opposition from officialdom prevented the realization.

[29] Figures according to Chen, *Qingdai junfei yanjiu*, pp. 297–298.

[30] Chen, *Qingdai junfei yanjiu*, p. 302.

The government under the Shunzhi emperor had also increased the salt tax. The region of Liang-Huai in 1678 yielded 47,000 *liang*, Changlu 44,000 *liang*, and Liang-Zhe (Zhejiang) 46,000 *liang*. By what means this extra tax was levied is not quite clear. Although there was an official rate of 25 *jin* per load (*yin*), with an extra surcharge of 0.025 *liang*, it seems that the money was raised according to need and not according to production or sale.[31] Kwan Man Bun speaks of the "predatory phase" in the history of the relation between the Changlu merchants and the government and quotes an increase of 45 per cent of the salt tax in 1645 and an additional levy of 0.05 *liang* per load in 1675.[32] In 1675 and 1677 the Kangxi emperor even announced rewards for officials delivering a higher amount of transit tax (*guanshui*) than the official quotas prescribed. Similarly officials could be promoted or were ranked higher on the waiting list for vacancies.

The precarious financial situation of the early Qing empire caused supervising secretary (*jishizhong*) Liu Yumo to bring forward the argument that the Shunzhi emperor had to build up military agro-colonies (*tuntian*) in the newly conquered territory in the southwest, especially in Sichuan. Liu said that the state income in 1652 was 14.859 million *liang*, while the expenditure was 15.734 million, 13 million of which served to finance war.[33] It was therefore necessary not only to collect supplementary taxes to cover the most urgent expenses for the military campaigns, but also to carry out land reclamation and a revision of land registers as a long-term means of increasing the state revenue. Although land reclamation appears to be only marginally related to war finance, this method provided the basis for a healthy income of the state because each registered tract of land was a source of taxation. The agro-colonies seem to have been successfully implemented in Hubei, Hunan and Shaanxi but failed to reach the projected revenue in Sichuan and the southwest. The government also refrained from founding agro-colonies in Taiwan.[34]

Another means of making the meagre state revenue more useful was to cut expenses, either by subtracting part of the officials' salaries, or by dismissing staff or eliminating posts, especially in local government. In 1650, for instance, the posts of 17 circuit intendants (*daoyuan*),

---

31 Chen, *Qingdai junfei yanjiu*, pp. 305–306.

32 Man Bun Kwan, *The Salt Merchants of Tianjin: State-Making and Civil Society in Late Imperial China* (Honolulu: University of Hawai'i Press, 2001), p. 36.

33 Liu Yumo, "Kenhuang xingtun shu", in *Huangchao jingshi wenbian* (Jindai Zhongguo shiliao congkan edition, 74; 731), juan 34, pp. 1234–1235.

34 Dai, "Yingyun shengxi", p. 6, footnote 20.

13 sub-prefectural magistrates (*tongzhi*) and 15 assistant sub-prefectural magistrates (*tongpan*) were cancelled. Two years later, 20 circuit inspectors (*dao yushi*) and 23 officials in the household administration of the heir apparent (*zhanshifu*), as well as 123 palace eunuchs (*taijian*) in bureaus of the Ministry of Works (*gongbu*), and 275 workers. In 1655 the government dismissed 198 granary workers in the provinces.[35] The local governments were furthermore granted fewer allowances for firewood, coal, clothing and furniture. Some officials saw even their salaries cut, part of which had to be delivered to the war chest as 'donated salary' (*juan feng*). In 1675 selected government institutions in Jiangsu had their funds cut. The governor-general had to donate his full nominal salary of 155 *liang*, was deprived of his 'seal paste and paper money' (*xinhong zhizhang yin*) to the amount of 288 *liang*, and of the stipends for secretaries to the amount of 120 *liang*. He did not receive any more money to employ gatemen, runners, sedan chair bearers and couriers. Similar cuts became effective for the office of the tax circuit intendant (*duliangdao*) of Jiangnan, the provincial administration commissioner (*buzhengshi*) of Jiangsu and the surveillance commissioner (*anchashi*) of that province. In 1680 officials in the province of Sichuan of the post of circuit intendant and higher were ordered to donate their full salary, those of a lower level half of it.[36] Local officials were deprived of their beadles, mounted constables, lantern bearers, and also of some—strictly seen not needed for administrative purposes—personnel like musicians, fan holders or umbrella carriers. The cost to employ all these people or to purchase seal paste and paper was to be covered by other means. T'ung-tsu Ch'ü has already written on the issue of the constant precarious financial situation of the district magistrates and says that, even under normal circumstances, "a magistrate's total salary was barely enough to pay for his private secretaries".[37] The solution to this problem was found in the *lougui* customary fee collected in the shape of various duties and charges imposed on the population. In the end, the cut of expenditure for the local government was refinanced by this customary fee and other surcharges.

An edict from 1678 recapitulates all the methods used in the past in order to cut spending and increase revenue for the sake of financing war, namely cutting expenses for the rations of officials in the courier stations,

---

[35] Figures according to documents in the *Qingshilu*, listed in Chen, *Qingdai junfei yanjiu*, p. 315.

[36] Figures see Chen, *Qingdai junfei yanjiu*, pp. 317, 319.

[37] Ch'ü, *Local Government in China*, p. 24.

using parts of the tax quota remaining in the provinces, transforming part of the tribute grain and objects to be delivered to the Imperial Household Department (*neiwufu*) into ready money, raising the salt tax and the real estate transfer tax, selling broker licenses and searching for unregistered farmland owned by state officials and those holding licenses for such.[38]

With the consolidation of the Qing regime and the professionalization and standardisation of its bureaucracy, a lot of these ad-hoc fundraising methods became obsolete or were transformed into a more regular form of levying charges from local officials or the people. The most important change was the shift towards institutionalized contribution campaigns in which officials or merchants 'voluntarily' donated funds or material. These institutionalized contributions are documented quite well, while the exploitation of individual officials continued, yet were only sporadically documented.

Chen Feng mentions in his discussion of Qing period war expenditure the share that state officials privately contributed to the cost in the shape of overspending that was not allowed to be brought to account.[39] Some examples can at least be quoted to demonstrate what conditions the Ministry of Revenue rated as essential to be accepted for proper accounting.

The brass used to cast cannons in Jinchuan proved not pure enough. The logistics managers were later able to provide brass with a better quality as well as expert craftsmen. Concerning the impure material it was ordered to find out who had supplied it, and this person was required to assume personally the transport cost to take this metal back to where it had come from.[40] An often-quoted item not allowed to bring to account was grass or hay for horses. From the beginning it was made clear that during campaigns horses should be pastured outside provincial territory and, as far as possible, not be fed with items that had to be carried from far away. The many admonitions in the edicts not to bring hay to account demonstrate that officials nevertheless tried to recover the cost for hay from the war chest.[41] This specification is also mentioned in the Jinchuan war expenditure regulations as well as in the *Junxu zeli* war expenditure code. Another, rather strange, instance of an official who had to cover losses from his salary is a case in which some Qing troops captured by the Jinchuan rebels were promised to be set free once ransom (euphemistically

38 *Pingding sanni fanglüe* (Siku quanshu edition), juan 36, fol. 4a (KX 17/3/guiwei).
39 Chen, *Qingdai junfei yanjiu*, pp. 226–227.
40 *Pingding Liang Jinchuan fanglüe*, juan 19, fol. 19a (QL 37/2/guiwei).
41 *Pingding Zhungar fanglüe, Qianbian*, juan 42, fol. 12a (QL 1/8/wuchen).

called 'reward') was paid.[42] It is very interesting to see that most officials made liable for back payment of overspending came from the local civilian bureaucracy and had the chance to cover these expenses in the one or other way by means of their function as bureaucrats in the tax administration, while military officials did not have access to such sources for compensation.

Last but not least, part of the war expenditure was financed in the frame of intra-provincial lending and liabilities, the clearance of which did not necessarily appear in the war expenditure account. Each province participating in the war pre-financed the baggage pay for the troops sent to the war theatre, organized their march to the camps, maintained the courier service, or provided transport animals, lead, saltpetre or other equipment. Theoretically such items were to be paid back by the war-waging province or the Ministry of Revenue, yet the examples at the end of the first chapter of this book have shown that it often took a decade or more before intra-provincial liabilities were finally settled, if they were balanced at all. This means that the statements of the share of each particular province in the expenditure for one single war are minimum figures that do not cover the whole share. The real share that each province contributed for war expenditure also includes the cost for items that were 'consumables' during the war like cannonballs, musket bullets, gunpowder and arrows or that had to be replaced due to wear and tear, like broken weapons, worn-out uniforms and boots, protective or winter clothing, tents or cooking pots. Expenditure for such items was—if not paid by the soldiers themselves—normally settled as current cost of the garrisons, for which the Ministry of Works had set up special regulations for the life expectancy (or 'amortisation period') of objects. Yet in fact, the depreciation of these objects was of course intensified during war, which makes their current cost hidden war expenditure. Cost was, in the end, funded by the province maintaining the respective garrison.

It should not be forgotten that the Qing armies were often supported by native auxiliary troops provided by allies of the Manchu dynasty. During the western campaigns the Mongol people of the Khalkha (Qalqa) supported the Qing troops regularly with horses, sheep and camels. The war period against the Dzungars between 1718 and 1739, for instance, cost the Khalkha about 4 million heads of cattle. In 1728 alone 3 million *liang* were spent for the purchase of beasts. In 1730 ten thousand Mongol

---

[42] *Pingding Liang Jinchuan fanglüe*, juan 32, fol. 10a (QL 37/6/jiashen).

warriors supported the Qing army. They were equipped and fed by their home tribe (*ayimaγ*) and not by the Qing government. Between 1753 and 1757 the four Khalkha tribes provided 150,000 horses, 221,000 camels, 22,300 cows and nearly half a million sheep for the Qianlong emperor's Dzungar campaign.[43] The share of the Mongols in the wars of the Qing empire, or those of the native kings in Eastern Tibet during the campaigns in Qinghai, Jinchuan and central Tibet, were in fact nothing else than tributes from the rulers of these peoples paid to the Qing court who in return were rewarded with presents as well as honorific titles. Yet in the end, the burden of these tributes was imposed on the shoulders of the common people that had to deliver their beasts and to provide labour, especially in the shape of underpaid military service as auxiliary troops or for the transportation of supplies.

Contributions

The share of institutionalized contributions for war expenditure is a theme that has only marginally been touched, even by experts like Xu Daling, Chen Feng and Lai Fushun.[44] Their expansion as a critical instrument for war-time fundraising during the nineteenth century has been studied by Elisabeth Kaske.[45] As contributions are a typical Chinese approach to generate income for the state, this matter shall be discussed here separately.

Contributions were, in other words, the sale of offices or at least that of brevet titles. This system originated in the custom of purchasing a college place in the Directorate of Education (*guozijian*) that administered the National University (*taixue*). Later on, officials actually occupying a position could buy an option for a higher ranking office. A district magistrate (*zhixian*) for example could purchase the option for the position of a prefect (*zhizhou*). This kind of contribution (*juanna*) was originally paid in rice, but later in money. It was not restricted in time yet subject to a fixed set of rules. The second type of contribution called *juanshu* was likewise strictly regulated but only used under conditions in which the government urgently needed money, mostly when natural disasters

---

43 Charles R. Bawden, *The Modern History of Mongolia* (London: Weidenfeld and Nicholson, 1968), p. 101. Veronika Veit, "Qalqa 1691 bis 1911", in *Die Mongolen: Beiträge zu ihrer Geschichte und Kultur*, ed. by Michael Weiers (Darmstadt: Wissenschaftliche Buchgesellschaft, 1986), pp. 453–454.

44 Chen, *Qingdai junfei yanjiu*, pp. 325–335. Lai, *Qianlong zhongyao zhanzheng*, pp. 377–381, 413–424.

45 Kaske, "Price of an Office". Kaske, "Fund-Raising Wars".

afflicted a region, large construction works like river conservancy works were undertaken or a military campaign was going on.[46] It was thus only a temporary affair. For this purpose, contribution campaigns were 'opened' (*kai juan*) or proclaimed that offered the opportunity to purchase a brevet title of an office or an option for a higher office. This kind of contribution was mostly paid in rice to support either the campaigning troops or the people afflicted by drought or inundation, but could also be paid in cash. Xu Daling interprets the *juanshu* contributions as a sub-form of *juanna* contributions.[47]

The government laid down the prices of brevet titles. The licence for a circuit intendant (*daoyuan*) for example cost 16,400 *liang*, that for a prefect (*zhifu*) 13,200 *liang*, and that for a district magistrate (*zhixian*) 4,620 *liang*.[48] Offices were for the first time sold through a *juanshu* campaign in the course of the early Kangxi reign, when the government was in dire need of money to finance the war against the Three Feudatories. During the Qianlong reign it became quite common that in the sphere of civilian offices options for posts up to circuit intendant could be purchased, in the military sphere up to the rank of assistant regional commander (*canjiang*). It was even admissible to purchase an option for offices in the central government up to the post of a director of a ministerial agency (*langzhong*). The purchase of options for military posts became sanctioned at the beginning of the Yongzheng reign.[49]

A special regulation of the assessment of the contribution provided that, when contributing rice, the government did not only count the market price of the donated grain (which ranged between 5 and 12 *liang* per *dan* of rice), but very generously calculated up to 25 *liang* per *dan* of rice.[50] This factor had been fixed during the first Jinchuan war but was later

---

[46] The role of private contributions for the granaries is described briefly in Pierre Étienne Will, *Bureaucratie et famine en Chine au 18e siècle* (Paris: Mouton: 1980), p. 170, footnote 32.

[47] Xu, *Qingdai juanna zhidu*, pp. 13–17.

[48] Figures according to Xu, *Qingdai juanna zhidu*, pp. 43–45. See also Lu Zijian, "Juanna: Qingdai de maiguan yujue zhidu", in *Wenshi zazhi* 1999/6, p. 75. Yang Zukai, "Qingdai guanli juanna zhidu qianxi yu sikao", in *Guizhou wenshi congkan* 1999/4, p. 38.

[49] Liang Yanbing and Liu Rong, "Qingdai de juanna zhidu", in *Lishi jiaoxue* 9 (1999), pp. 50–53. I was not able to find out the exact date stating when military offices had been purchased for the first time (presumably by lower-ranking officers aspiring to a higher rank). Cases in point are military candidates asking to be promoted to a rank after delivering a contribution: Holders of the military *jinshi* degree (a degree conferred after passing the military examinations) were promoted to the rank of *shoubei*, holders of the *ju*[*ren*] degree to that of *qianzong*. *Qingshilu, Shizong shilu*, juan 5, fol. 3b–5a (YZ 1/3/xinsi).

[50] *Qingshigao*, juan 112, p. 3235.

criticized by the emperor as being too much.[51] The original proposal of the Sichuan governor Jishan had been to value each *dan* of rice with 30 *liang*, which was up to six times the market price and 25 times the official price for grain (1.2 *liang* per *dan*). The emperor therefore harshly criticized this method—although he finally accepted the conversion factor of 25 *liang*. His scepticism turned out to be justified when after the end of the war it was discovered that many officials had chosen to have their daily rations paid out in rice instead of cash. The officials then stored the rice in order to donate it during a contribution campaign, which would save them a lot of money because the value of each *dan* of rice would then increase at least sixfold.[52] Thus, if a contributor provided 1,000 *dan* of rice (with a value of between 500 and 1,200 *liang*) he could obtain a licence worth 25,000 *liang*. To tell it in a different way, when selling the option for the post of a district magistrate, the government did not obtain rice with a market value of 4,620 *liang*, as the statutes for the *juanshu* contributions had fixed, but of only about 185 *liang*. When troops in the field wanted cash instead of rice for daily rations—which was not always allowed—the normal conversion rate was 4 to 6 *liang* per *dan*. In order to forestall this illegal practice governor-general Zhang Guangsi suggested enforcing the correct distribution of money, not of rice. Governor Jishan and his provincial administration commissioner Cang Dejun were dismissed, yet the non-distribution of rice deprived the troops of part of their supplies, so that they in some camps ran out of food and the handing out of rice was allowed again, yet at a higher price of 6 *liang* per *dan*, which somewhat reduced the attractivity of this kind of 'arbitrage' business.[53]

Although the contribution scheme was "hardly an instrument of choice for Chinese governments", as R. Kent Guy says, it was fairly commonly used, especially when the government treasury was under financial stress, e.g. in military circumstances.[54] Guidelines or statutes for the procedures of *juanshu* contributions were laid out based on the experiences made in many campaigns during the early Qing period. Such were the "Rules from Jiangxi, Fujian and Hu-Guang (Hubei and Hunan)" (1674, war against Wu Sangui), the "Rules for the contribution of grain and fodder" (1695, the war against Galdan), the "Rules for the logistics of the northwestern

---

[51] *Jinchuan dang* 37/IV/00184 (QL 37/11/12).

[52] *Gaozong shilu*, juan 309, fol. 57b–59b (QL 13/2/jiashen). *Jinchuan dang* 37/IV/00184 (QL 37/11/12).

[53] Narrated in Lai, *Qianlong zhongyao zhanzheng*, pp. 378–379.

[54] Guy, *Governors and their Provinces*, p. 219.

[campaigns]" (1715), the "Rules for grain transport to the front" (1722, conquest of Qinghai), "A'rtai's rules for grain transport" (1724, Qinghai campaign), the "Rules for the transport of grain, as arranged by the Ministry of Revenue" (1734, Miao rebellions), the "Rules for the transport of grain to Jinchuan" (1748, first Jinchuan campaign), the "Rules for the [grain] transport in Sichuan" (1774, second Jinchuan campaign), or the "Rules for arranging the reconstruction in Sichuan und Hunan" (1798, White Lotus rebellion). From the titles of these regulations it can be seen that the contributors took over a part of the logistics by purchasing and transporting grain directly to the camps or at least to a hub in the logistics lines. This did not only indirectly provide funds to the government, but also relieved the logistics managers of part of their work.

While *juanshu* contributions provided an excellent opportunity for social and eventually financial advancement of most state officials or private persons, they had a very negative aspect for those who were forced to donate money, first and foremost the prosperous salt merchant associations. That for them the *juanshu* contributions were nothing but an irregular and unofficial tax in a procedure called *baoxiao* 'to announce an offering' can also be seen from the fact that special contribution rates existed for certain regions. In 1732 the merchants from the salt administration region Changlu (Tianjin) made a first contribution of 100,000 *liang* to the imperial military expenditure and from then on, whenever there was a greater military expedition, the salt merchants of Changlu, Shandong, the Huai region and the Liang-Zhe region contributed large sums at a certain rate according to their estimated wealth. During the second Jinchuan war, in 1773, the Huai salt merchants offered a contribution of 4 million *liang*, the Liang-Zhe salt merchants paid 1 million *liang*, the Changlu salt merchants 600,000, and the Shandong merchants 300,000.[55] Yet apart from the extraordinary contributions during wars (the 'donations'), there were also regular, project-related contributions for disaster relief or river conservancy works, which likewise could amount to more than 1 million *liang* for the Huai merchants and up to several hundred thousand *liang* for the other salt merchants. Although the *juanshu* campaigns thus provided a lot of possibilities for social and financial advancement, it is a fact, especially for the salt merchants as well as the bankers in Shanxi, that the campaigns also served to fleece them. Chen Feng has compiled a list from which

55 Figures according to Yanwushu (Salt administration bureau), ed., *Zhongguo yanzheng yange shi: Changlu*, in *Jindai Zhongguo shiliao congkan*, 64, vol. 637, p. 42.

it can be seen that during the Shunzhi reign the Liang-Huai merchants contributed 300,000 *liang*, during the Kangxi reign all salt merchants about 350,000 *liang*, of which 200,000 alone were used for the campaigns against the Dzungars. About the same sum was raised during the much shorter Jiaqing reign. The longest contribution campaign took place during the war against the White Lotus rebellion with a length of five years, in which nearly 18 million *liang* were collected.[56] Elsewhere Chen speaks of no less than 30 million *liang*, or, including other contributions during that time, of even 70 million.[57] During the whole, sixty-year-long Qianlong reign wartime contributions accrued to a sum of 21.9 million *liang* or even more. Salt merchants were even fleeced to pay peacetime stipends for the army, like donations for weddings and funerals.[58] Kwan Man Bun sees the increasing role of contributions as a "price of their prosperity". In the early seventeenth century the Imperial Household Department had regularly provided the salt merchants with credits, a tool that enabled them to make a fairly profitable business and at the same time allowed the government to skim off a sufficient amount of taxes. In the eighteenth century the donations can therefore be seen as a kind of back-payment of these earlier loans including interest made by the government.[59]

Contributions under the Qing were introduced as early as during the Shunzhi reign. There is, for instance, a document showing that Banner troops, from generals down to common soldiers, were obliged to contribute for disaster relief.[60] This kind of fundraising only became a regular practice during the war against the Three Feudatories and in the western campaigns during the Kangxi reign. In 1696 state officials donated 210,000 *liang*. A three months' campaign in 1715 yielded for instance 4.4 million *liang*, contributed by 16,787 persons. In the years thereafter, a further sum of 10 million *liang* was raised. The governor of Guangxi alone contributed 4,000 *liang*, a textile producer 3,000, the governor of Yunnan even 22,000 and the governor of Shandong 30,000.[61] For the western campaigns during the Yongzheng reign, contribution campaigns were likewise opened that were at that time still called "temporary *juanna* contributions".[62]

---

56 Chen, *Qingdai junfei yanjiu*, pp. 332–333.

57 Figures quoted in Chen, *Qingdai junfei yanjiu*, p. 330.

58 Dai, "Yingyun shengxi", p. 47.

59 Kwan, *Salt Merchants of Tianjin*, p. 43.

60 It is not known if they were later rewarded with promotions. *Qingshilu, Shizu shilu*, juan 89, fol. 9a (SZ 12/2/jisi).

61 Figures quoted in Chen, *Qingdai junfei yanjiu*, p. 330.

62 *Pingding Zhungar fanglüe, Qianbian*, juan 14, fol. 27a (YZ 2/9/guimao).

Local officials of the province of Gansu contributed carts and fodder for horses.[63] General E'rtai especially recommended that contributions were delivered in grain and not in money, so that the army might immediately be supplied.[64] The governor of Shanxi also suggested deducting a certain amount of money from the *yanglian* 'anti-corruption' salary of the provincial officials as contributions, in order to cover expenses for salt, tea or medicine.[65]

Although contributions played an ever-increasing part in the finance of wars, the rhetoric of the emperors feigned a negative stance towards this financing tool, as can be seen in the assertion of the Yongzheng emperor quoted at the beginning of this chapter, that already his father had suspected institutionalized contributions as "not a praiseworthy matter",[66] or in the Qianlong emperor's admonition to his successors to never again make use of contribution campaigns.[67] The caution with which the Qing rulers regarded the contributions is justified by a considerable number of embezzlement cases, the most notorious one being an affair exposed in 1781. Cash contributed for relief payments found its way into the private pockets of a large number of officials. Money, "contributed to secure private ends, [was] being recycled through public officials to individuals in need".[68] This *cause célèbre* demonstrates how difficult it often was to separate private funds from those of the public, not only on the highest level of government (with the funds of the Imperial Household as the private money of the dynastic family), but also on a local level.

In his early years the Qianlong emperor had therefore attempted to make do without contribution campaigns, yet already during the first Jinchuan campaign, begun in the thirteenth year of his reign, it was decided to raise funds again by opening such a campaign. It proved to be a problem that at the same time a contribution campaign was opened to bring disaster relief in the province of Shandong. In order to forestall profiteering persons from Sichuan were not allowed to take part in it, instead they

63 *Pingding Zhungar fanglüe, Qianbian*, juan 22, fol. 5b–6a (YZ 9/2/wuwu, and 4/gengzi).

64 *Pingding Zhungar fanglüe, Qianbian*, juan 33, fol. 2a (YZ 10/11/gengzi).

65 *Pingding Zhungar fanglüe, Zhengbian*, juan 56, fol. 19a (QL 23/6/dingsi). The *yanglian* pay (literally 'nourishing incorruptibility') was introduced in 1723 as an addition to the nominal pay of officials (*fengxiang*), in order to forestall excessive corruption. Its amount was substantially higher than the regular pay.

66 *Pingding Zhungar fanglüe, Qianbian*, juan 14, fol. 27a (YZ 2/9/guimao).

67 *Gaozong shilu*, juan 1441, fol. 3a–5a (QL 58/11/yiyou).

68 Guy, *Governors and their Provinces*, p. 219.

had to deliver their donations to the contributions bureau (*juannaju*) in Sichuan. The statutes for contributions were at that time far from perfect and often had to be adapted to circumstances. The emperor was even very reluctant to open a new contribution campaign for military matters at the beginning of the second Jinchuan war. The reasons for his hesitating stance and the final outcome of deliberations will be seen later. Although it is often said, even by Lai Fushun himself, that contribution campaigns were carried out only during the two Jinchuan wars,[69] it can be proved that practically all wars were co-financed in this way. Contributions had become such a helpful tool in extraordinary situations like wars and natural disasters that they were practically used for a wide variety of issues for which the government needed funds.[70]

Lai Fushun provides a detailed list of contributions for the great wars during the Qianlong reign.[71] According to this list, 300,000 *liang* of money and 5,000 *dan* of rice were contributed during the first Jinchuan campaign by salt merchants of Jiangxi and Guangdong, high officials in Jiangxi and the local gentry in Sichuan. In many cases, the donors had to borrow money from the treasury of the administration circuit that was to be paid back later. The governor of Jiangxi alone paid 10,000, the provincial administration commissioner 7,000 *liang*. For the first Dzungar campaign merchants from Fujian, Zhejiang, Jiangxi and the salt administration zones of Liang-Zhe and Huai-Hui (Liang-Huai) contributed money to an amount of 1.9 million *liang*, while Mongol allies contributed thousands of horses and camels. In this case, too, the merchant associations had to borrow money from the government, with the stipulation to pay it back over the next few years. 1.2 million *liang* came from the salt merchants of Liang-Zhe and merchants from Zhejiang for the war against the Muslim insurgents in 1757, and again, local allies provided huge numbers of transport animals as well as cattle and sheep for consumption. The Taiwan campaign is in many respects an interesting case differing from other wars that were exclusively fought on land. Salt merchants from Liang-Huai, Changlu, Shandong and Liang-Zhe contributed 2.9 million *liang*, merchants from Hedong (Shanxi) donated 300,000 liang, merchants from Henan brought in a volume of 20,000 *dan* (2,070 m^3) of black beans, and overseas merchants from Guangdong supplied 500,000 *liang*. Compared

69 Lai, *Qianlong zhongyao zhanzheng*, p. 201.

70 During the Xianfeng reign contributions were also used to put newly issued banknotes and heavy coins into circulation.

71 Lai, *Qianlong zhongyao zhanzheng*, pp. 412–424.

with the other campaigns, the total sum of contributions for this war appears to be rather large. The second Jinchuan campaign was, according to the documents collected by Lai Fushun, supported by 7.7 *million* liang.[72] This large amount will be discussed later. The contributions for the Myanmar campaigns were relatively modest. Lai lists an amount of 1 million *liang* contributed by the salt merchants of Liang-Huai. For the second Gurkha campaign the salt merchants of Liang-Huai, Changlu, Shandong and Liang-Zhe, as well as overseas merchants and salt merchants from Guangdong, contributed 4.1 million *liang*. Horses and mules were bought and donated by the Mongols of Qinghai, and several high government officials. The governor-general of Shaan-Gan (Shaanxi and Gansu), Lebao, for instance, financed 500 horses (with a value of about 5,000 *liang*), and the governor of Shanxi 1,000 mules. Huiling, governor-general of Sichuan, and some circuit intendants provided 3 *liang* of silver for each of the Banner troops passing Chengdu. For this campaign 6 million *liang* had been allotted, as was seen before, of which 4 million originated in contributions that proved, in the end, not necessary. This is quite a strange phenomenon because the emperor normally did not want contribution campaigns to be opened prematurely or without urgent need, although the official explanation was to finance the 'victorious return' of the troops with the help of contributions, and not the campaign itself. At least the contributions by officials might have been motivated by liabilities towards the government on their part, either stemming from the campaign directly, caused by overspending, or from an earlier time. Similar to this, the contributions by merchants can be seen as liabilities of this group of persons towards the government in the shape of taxes. The regular character of contributions imposed on merchants might prove that they were just another form of taxation, disguised as a voluntary donation to the cause of the nation.

War Finance in the Second Jinchuan Campaign

In the second half of the Qianlong reign the experiences of a series of military campaigns resulted in a very smooth method to finance war, even if it was such a long and expensive one as the second Jinchuan campaign. It is therefore all the more interesting to examine how the Qing government financed this war. Compared to the many ad-hoc methods of fundraising

[72] Lai, *Qianlong zhongyao zhanzheng*, pp. 419–421.

still used under the Shunzhi and Kangxi emperors, the campaigns of the Qianlong emperor appear as nearly perfect bureaucratic models of war finance.

The Origin and Transfer of Funds

At the end of the *Jinchuan junxu li'an*, the regulations for war expenditure for the second Jinchuan campaign, a general accounting for the whole campaign is to be found.[73] The total sum in the books of the Ministry of Revenue was more than 64,109,800 *liang*. 1.1 million of this originated in the shape of regular and miscellaneous taxes in the provincial treasury of Sichuan; the same amount was spent from among the finances in hand held ready for military expenditure in the treasuries of the garrisons of Songpan and Dajianlu;[74] 250,000 *liang* came from the treasury of various administration circuits (*daoku*);[75] and more than 59,800 *liang* were provided from the tea tax (*chaxi*) collected in Sichuan.[76] During the first months of the campaign cost was still covered by the provincial funds of Sichuan allotted for unexpected war expenditure ('barbarian affairs'), and during that period of time the baggage pay for the troops was covered by the tea tax.[77] It was thus money owned as tax revenue by the provincial treasury that the governor-general had used for the campaign.[78]

---

[73] *Jinchuan junxu li'an*, juan 2, fol. 177a–187a.

[74] Songpan is situated north of the Jinchuan area, Dajianlu (modern Kangding) to the south. Both were important garrison posts protecting major roads: Songpan the one from Sichuan to the province of Gansu, and Dajianlu that to Tibet. Why both garrisons possessed such large sums of money is not really clear. Probably those funds had been allocated to the two frontier garrisons to be instantly ready for troops suppressing rebellious mountain tribes.

[75] The *Jinchuan junxu li'an* does not speak of any particular circuit, yet the region west and northwest of Chengdu was administered by the Cheng-Mian-Long-Mao circuit, which comprised the prefectures of Chengdu, Mianzhou, Long'an (modern Pingwu), and Maozhou (modern Maowen), while the south and southwest of Chengdu, like the prefectures of Qiongzhou (modern Qionglai), Yazhou (modern Ya'an), Meizhou (modern Meishan) and Jiading (modern Leshan), were under the jurisdiction of the Jianchang circuit. The respective money probably came from both those circuits, or even from other circuits in the province of Sichuan. The circuit treasuries contained the salary for the military garrisons in that region. Compare Huang Tianhua, *Zhongguo caizheng shigang* (Shanghai: Shanghai caijing daxue chubanshe, 1999), p. 335.

[76] See also *Jinchuan dang* 36/III/00025 (QL 36/8/8).

[77] *Jinchuan dang* 36/III/00025 (QL 36/8/8). *Pingding Liang Jinchuan fanglüe*, juan 6, fol. 18a (QL 36/8/bingzi). This was quite usual in Sichuan, like in QL 32 (1767), when a period of peace in Jinchuan made it possible to allot the tea tax fund for the repair of the city wall of Chengdu, for example, *Pingding Liang Jinchuan fanglüe*, juan 5, fol. 11a (QL 32/2/wushen).

[78] *Jinchuan junxu li'an*, juan 1, fol. 1a.

Table 7. Origin of the money disbursed for the second Jinchuan campaign.

origin		sum (million *liang*)
Sichuan		
provincial treasury		1.1000
military budget of the garrisons in Songpan and Dajianlu		1.1000
circuit treasuries		0.2500
tea tax		0.0598
sub-total		2.5098
Ministry of Revenue		29.0000
various provinces and salt administration units		
Guangxi		1.2000
Guangdong		1.8000
Shandong		0.8000
Shanxi		0.9000
Hubei		1.6000
Hunan		3.1000
Shaanxi		2.5000
Henan		3.6000
Jiangsu		3.8100
Anhui		2.3000
Liang-Huai		3.4000
Zhejiang		0.4900
sub-total		25.5000
salt merchants		
Liang-Huai[a]		2.0000
Changlu		0.9000
Zhejiang		1.0000
Shanxi		1.1000
Guangdong		0.2000
Guangxi		0.4000
sub-total		5.6000
Miscellaneous	sum (*liang*)	
booty from Gara'i	5,869	
transport surcharge (*yuping*) from Zhejiang	7,350	
repayments in exchange for rice	118,525	
various repayments	9	
sub-total	131,754	
total sum (*liang*)		62,741,554

[a] Of the promised 4 million *liang*, only 2 million were paid until the end of the contribution campaign.

Source: *Jinchuan junxu li'an*, juan 2, fol. 177a–187a.

The treasury of the Ministry of Revenue itself provided 29 million *liang*; another 25.5 million *liang* were made available by various provinces (other than Sichuan); and 5.6 million *liang* were disbursed by salt merchants as contributions, at least until the end of 1773, when the contribution campaign for the second Jinchuan war was ended.

According to these statements, a total sum of 62.7 million *liang* was the 'income' of the Ministry of Revenue. Compared to the 64 million *liang* mentioned above, some 1.5 million *liang* are missing.

In the list, another, not yet mentioned means to finance war is included, namely booty worth 5,869 *liang*. The Qing troops apparently regularly looted the conquered monasteries and had to make lists of the objects found.[79]

Another 7,350 *liang* were collected from the province of Zhejiang in the form of transport surcharges for the transport of silver ingots. A sum of 118,525 *liang* were repaid by the officials in the war logistics bureau (*junxuju*), which first had spent this sum to pay labourers but later switched over from money to rice portions because this payment was meant for a period of non-productivity on the way back to the logistics stations without carrying a load. A tiny sum of 9.375 *liang* (with the value of about one horse) came from "miscellaneous repayments". These various sums (amounting to 131,754 *liang*), added to the 62.6 million *liang* directly paid out by government agencies and salt entrepreneurs, made for a total income of the war chest of 62,741,554 *liang*. The account book of the *Jinchuan junxu li'an*, nevertheless, only gives a total sum of 61,731,754 *liang*—about exactly one million *liang* short of the sum total of all sources of income.[80] It is not possible to trace the origin of this error. The difference to the 64 million *liang* is likewise not explained.

---

[79] *Jinchuan dang* 36/III/00099 (QL 36/9/20), like Buddha figures and other valuable objects found in the castles of Greater and Lesser Jinchuan. The native auxiliary troops usually were allowed to plunder the less valuable objects, like grain and cattle, in order to reward them for their willingness to fight against their neighbours. The booty taken from the castle of Senggedzung, for example, consisted of weapons (muskets, cannons, swords, spears), gunpowder, bullets, cattle (sheep, cows), grain and other eatables, see *Jinchuan dang* 37/IV/00283 (QL 37/12/3). It was even the custom to seize the private property of some of the rebel leaders, like in the case of Damba Wodzar whose property of 156 *liang* was confiscated by the imperial troops, see *Jinchuan dang* 41/II/00043 (no date).

[80] *Jinchuan junxu li'an*, juan 2, fol. 177b–178a.

Source: *Pingding Liang Jinchuan desheng tu* (copper plate engravings), Staatsbibliothek Berlin, Preußischer Kulturbesitz, Signatur Libri sin. 1603 (Sequence Qurman: II 5, Sammlung: II).

Figure 3. Presentation of booty and heads of enemies after the conquest of Ripang.

It might be that part of the difference was bridged by the *pingyu* surcharge levied in other parts of the empire (and not only in Zhejiang, as the list above pretends). The *pingyu* surcharge had, although its origins went back to the difference between the official weights in the capital and those in the provinces, for a long time served to finance part of the income of state officials. It was a regular fund to finance the *yanglian* 'anti-corruption' pay, an abundant surplus payment to the nominal salary.[81] In Sichuan, for instance, an amount of three *liang* per official and year was to be paid from the *pingyu* surcharge collected for the provincial treasury.[82] It was added as surcharge to the normal tax and constituted between 0.6 per cent and 2.5 per cent of the basic land tax. During the Jinchuan war, the *pingyu* surcharge had the following effect: The regular part of the military budget (*zhengxiang*) served to finance the most expensive items, like the baggage pay of the soldiers (*xingzhuang*) and their monthly 'salted-vegetable pay' (*yancai*) as well as cost for transporting grain and military equipment and the maintenance of transport and courier stations. Smaller, 'miscellaneous' items (*zakuan*), like the construction of roads and bridges, building sheds and stables, medicine, and all kinds of material for bureaucratic needs, were to be paid from the money coming in through the *pingyu* surcharge. Yet the problem was that in Sichuan it was not common to use this kind of tax for funding wars, as had been usual during the western campaigns in the province of Gansu. Therefore governor-general Wenshou suggested using of the *pingyu* surcharge to finance the war in Sichuan, too. In Gansu the procedure was to collect the tax and to use it for financing one per cent of all military expenditure, but this amount seemed too high, so that Wenshou suggested another procedure: half a per cent of the expenditure for all items were to be used to finance the 'miscellaneous items'. This part of the budget was then earmarked for these items of the military expenditure as well as for payments to private transport entrepreneurs. The decision to finance those items by the *pingyu* surcharge was made on QL 37/11/12 (Dec 6, 1772), but it could hardly have been the case that all expenditure for the private transport entrepreneurs (amounting to probably 25 million *liang*) could have been financed by the *pingyu* surcharge from the province of Gansu, and therefore an additional paragraph in the *Jinchuan junxu li'an* says that it only served to finance the additional clothing for the soldiers in winter, the pay for labourers, the construction of roads and bridges, and to buy horses for the stations. Yet even for financing all those matters the income of the Sichuan *pingyu* surcharge was insufficient.[83] The final account gives the figure of 134,759 *liang*, and there it is stated that the *pingyu* surcharge was only siphoned off from the transport cost for the grain and for the

81 To the *yanglian* pay, see Ch'ü, *Local Government in China*, pp. 22–24.
82 *Pingding Liang Jinchuan fanglüe*, juan 54, fol. 6b (QL 38/3/xinhai).
83 *Jinchuan junxu li'an*, juan 2, fol. 138a–139a.

Table 8. Items paid for by the *pingyu* surcharge.

item	sum (in *liang*)
rewards for native kings after the war	11,970
medicine	4,785
stationery	25,289
monthly expenditure for secretaries	79,444
escorting prisoners	8,041
construction of bridges and roads, boatbuilding	4,850
indemnities for drowned labourers	479
sum	*134,858*

Source: *Jinchuan junxu li'an*, juan 2, fol. 186a–187a.

purchase of other equipment and horses.[84] The same procedure is reflected in the war expenditure regulations *Junxu zeli*, where it is said that the *pingyu* cannot be deducted from the baggage pay or the 'salted-vegetable pay', but only from expenditure incurred for the purchase of equipment or from the grain transported by private entrepreneurs.[85]

The *pingyu* surcharge was also imposed on the rice transported to the camps, where it was common that in the grain stations of the logistics routes the responsible grain official did not only surcharge a transport loss of one *sheng* per *dan* of rice (a hundredth), but another *sheng* of rice to finance the construction of roads and to feed labourers. This second surcharge had to be brought to account as *pingyu* surcharge, and to be kept separate from the main account.[86] How much income the generals had obtained by levying the *pingyu* surcharge and for which purposes it was spent, has been summarized in a list,[87] which includes items normally not mentioned as those for which the surcharge was intended to be used, especially rewards for the native chieftains that had sent auxiliary troops, or indemnities paid out to the families of labourers that had drowned.

As a résumé it has to be concluded that according to the data given in the *Jinchuan junxu li'an*, the Ministry of Revenue or the central government financed 46 per cent of all military expenditure. 40 per cent were contributed by various provinces and salt administration units; if the province of Sichuan is included, which spent 4 per cent of all the expenditure,

84 *Jinchuan junxu li'an*, juan 2, fol. 186a.
85 *Hubu junxu zeli*, juan 9, fol. 9b–10a.
86 *Pingding Liang Jinchuan fanglüe*, juan 101, fol. 6b (QL 40/7/jisi).
87 *Jinchuan junxu li'an*, juan 2, fol. 186a–187a.

the provincial governments had spent about the same share as the central government. Almost 9 per cent of the total expenditure was financed by entrepreneurs through their contributions.

The responsibility for financing a war like the Jinchuan campaign was shared by different administrative levels of the government. This can best be seen with the financial transactions between the central government and its sub-divisions—in the first instance the Ministry of Revenue—and the local governments in the provinces. The largest burden in the field of organization of course rested on the shoulders of the province of Sichuan which was not only the place from where all troops and the objects they required had to be taken towards the mountains: the governor-general of Sichuan also played a crucial role in the command structure and the administration commissioner (*buzhengshi*) took care of the logistics and accounting.

At the beginning of the campaign a considerable part of the war chest was fed by the provincial treasury of Sichuan. Only when it ran out of money, new funds were transferred from the Ministry of Revenue or from other provinces. In many cases and especially in the first phase of the war the emperor ordered the Ministry of Revenue to investigate how much money could be forked out from the provincial treasuries. Apparently the finances of the central state treasury in the Ministry of Revenue were not to be touched as far as possible. For example, instead of being paid by the Ministry of Revenue the sum of 3 million *liang* to be sent to Sichuan as the first financial support on QL 36/11/15 (Dec 20, 1771), was composed from various taxes that had been levied in different provinces. Although the *Pingding Liang Jinchuan fanglüe*, the official annals of the war, do not report all transfers, it is by means of other documents (with few exceptions and some doubts) nevertheless possible to construct an appropriate overview of a great part of the fifteen transfers from the treasury of the Ministry of Revenue or from the budgets of various provinces. In some documents, sub-totals are given that are very helpful for the reconstruction of transfers, because those sums allow to check if an amount in-between is missing in the documents, as e.g. in the year's span between QL 38/6 (Jul–Aug 1773) and QL 39/5 (Jun–Jul 1774).

The first amount are the 3 million *liang* disbursed on QL 36/11/15 (Dec 20, 1771), for which the Ministry of Revenue had made a proposal as to what province was to provide the respective smaller amount of money to be sent to the destination in Sichuan. A large part of the sum came from the salt taxes in the provinces of Guangxi and Guangdong. According to the *Huangchao wenxian tongkao* "General history of

Table 9. Monetary transfers to Sichuan in order to fund the second Jinchuan war (in *liang*).

date[a]	origin	amount
QL 36/11/15	various provinces near Sichuan:	3,000,000
	Hubei	(300,000)
	Hunan	(300,000)
	Guangdong	(600,000)
	Guangdong salt tax	(600,000)
	Guangxi salt tax	(1,200,000)
QL 37/1/15	*intermediate sum*	*3,000,000*
QL 37/2/27	(Ministry of Revenue)	3,000,000
QL 37/4/25, QL 37/5/8, QL 37/5/22, QL 37/6/24	*intermediate sum*	*6,000,000*
QL 37/6/24	Hu-Guang	2,000,000
QL 37/7/5	Ministry of Revenue	1,000,000
QL 37/10/4, idem *Jinchuan dang*	*intermediate sum*	*9,000,000*
QL 37/10/4	various provinces near Sichuan	2,000,000
QL 37/10/27, *Jinchuan dang* QL 37/10/29	(Ministry of Revenue)	3,000,000
QL 37/10/27, QL 37/11/12, *Jinchuan dang* QL 37/12/13	*intermediate sum*	*14,000,000*
QL 38/1/4	Ministry of Revenue	5,000,000
QL 38/i3/3	*intermediate sum*	*19,000,000*
QL 38/i3/3[b]	various provinces:	5,000,000
	Henan	(900,000)
	Anhui	(900,000)
	Jiangsu	(1,250,000)
	Jiangning land-taxes	(260,000)
	Hubei	(100,000)
	Hunan	(500,000)
	Liang-Huai salt tax	(600,000)
	Zhejiang salt tax	(490,000)
QL 38/5/9, QL 38/5/13	*intermediate sum*	*24,000,000*
QL 38/5/13	Ministry of Revenue	5,000,000
QL 38/6/3, QL 38/8/21, *Jinchuan dang* QL 38/7/20	*intermediate sum*	*29,000,000*
(No respective document)	(contributions)[c]	[5,000,000]
QL 39/5/6	*intermediate sum*	*34,000,000*
QL 39/5/6	various provinces	5,000,000
Jinchuan dang QL 39/7/20	*intermediate sum*	*>39,000,000*
QL 39/9/>7,[d] *Jinchuan dang* QL 39/9/27	Ministry of Revenue	4,000,000
QL 39/11/23, idem *Jinchuan dang*, and QL 39/12/17	Ministry of Revenue	5,000,000

Table 9 (*cont.*)

date[a]	origin	amount
QL 40/1/22, idem *Jinchuan dang*	Ministry of Revenue and provinces	5,000,000
	Jiangxi, Guangdong and Gansu[e]	3,370,000
	Ministry	1,630,000
QL 40/2/21,[f] *Jinchuan dang*	*intermediate sum*	*>40,000,000*
QL 40/2/9		
QL 40/5/26	Ministry of Revenue	5,000,000
QL 40/9/29[g]	Ministry of Revenue and provinces:	3,000,000
	Henan	(500,000)
	Shandong	(500,000)
	Shanxi	(800,000)
	Ministry	(1,200,000)
QL 40/11/24	Ministry of Revenue	3,000,000
	total sum	*64,000,000*

a) Date of issue of the respective document as recorded in *Qingshilu*.
b) Details according to *Pingding Liang Jinchuan fanglüe*, juan 55.
c) There is no document proving that the missing 5 million *liang* (to reach the sub-total of 39 million *liang*) were disbursed from the ministerial or provincial treasuries. Between mid-QL 38 and mid-QL39 no large sums were dispatched to Sichuan apart from the contributions by salt merchants.
d) According to QL 39/12/20, QL 39/12/26 and QL 40/1/7. There is a document dated QL 39/9/7 that simply orders that 2 to 3 million *liang* be dispatched. In the end 4 million *liang* were sent in the 9th month, as the documents of the 12th month state.
e) According to *Junjichu lufu zouzhe*, reel 35, no. 613. The balance could have been paid by the Ministry of Revenue.
f) *Junjichu lufu zouzhe*, reel 590, no. 2461. The sum given in this document should be much higher. The author probably refers to money that had already arrived in Sichuan and which could still be less than 50 million *liang*.
g) Details according to *Junjichu lufu zouzhe*, reel 35, no. 792.

Source: *Qingshilu, Gaozong shilu* and *Jinchuan dang*, individual dates of documents; *Pingding Liang Jinchuan fanglüe, Junjichu lufu zouzhe*, see individual footnotes above.

institutions and critical examination of documents and studies of Our August Dynasty" the official annual salt tax rate of the two provinces was about 47,000 *liang* each.[88] When both provinces had to pay a sum that was the twelve-fold or even the twenty-five-fold of that rate we learn much about the difference between the official tax rate and the real taxes levied. The province of Guangdong additionally had to pay 600,000 *liang* from its salt tax income—which constitutes half the annual land-tax

88 *Huangchao wenxian tongkao*, juan 40, *Guoyong* 2, pp. 5525–5228.

rate as listed in the *Huangchao wenxian tongkao*, and a fourth of the real land-tax income of the province, according to Yeh-chien Wang.[89] The two provinces of Hubei and Hunan with their mid-range tax income only had to supply half that sum, i.e. 300,000 *liang*.

The second case where it can be learnt about the origin of the money sent to Sichuan is a document dated QL 38/13/3 (Apr 24, 1773). Again, some provinces like Henan, Anhui, Jiangsu, Hunan and Hubei had to send part of their land-tax income to Sichuan, the largest sum of 1.25 million *liang* being paid by the rich province of Jiangsu whose official land-tax rate was somewhat more than 3 million *liang* annually according to the *Huangchao wenxian kongkao*, while the real land-tax income was about 8.5 million *liang*, according to Yeh-chien Wang's estimations. Even the relatively poor province of Anhui had to disburse 900,000 *liang*, which is about half the total annual tax rate of that province of nearly 1.8 million *liang*. And again, the salt tax income of some provinces was so abundant that Zhejiang province and the regions under the Liang-Huai salt administration (located in the northern part of Jiangsu) had to disburse about half a million *liang* each.

On QL 37/6/24 (Jul 24, 1772) the money coming from the two provinces of Hubei and Hunan (Hu-Guang) seemed to be so lavish that they were able to disburse 2 million *liang* for the war.[90]

For the 3 million *liang* transferred in QL 40/9 (Oct 1775, exact date of respective edict not known) details about the origin of this sum can be found in a memorial among the State Council copies (*Junjichu lufu zouzhe*) which states that of the 3 million *liang*—whose purpose is defined as reconstruction (*shanhou*) fund—more than one third came from the Ministry itself while the rest was paid by the provinces Henan, Shandong and Shanxi.[91]

While these documents reveal what amounts of money came from what province, there are other sums having come from various provinces about which there is no concrete statement as to their exact origin, nor what types of source (land-tax, salt-tax or probably transit taxes?) were used to finance the war. Although the richest provinces being best able to pay the largest sums for the war were located in the eastern parts of Qing

---

89 Wang Yeh-chien, *An Estimate of the Land-Tax Collection in China, 1753 and 1908* (Cambridge: Harvard University Press, 1973), Table 27.

90 *Gaozong shilu*, juan 911, fol. 21a (QL 37/6/wuzi).

91 *Junjichu lufu zouzhe*, reel 35, no. 792.

China, the government tried to create a balance when imposing the duty to support Sichuan province. In two documents it is therefore said that the 2 (resp. 3) million *liang* to be sent were to come from the provinces located near Sichuan.[92] This means that that money came from the provinces of Shaanxi, Shanxi, Hubei, Hunan, Guizhou and/or Yunnan. Because Hubei and Hunan had already contributed considerable sums in the year before and the following spring, those two provinces might this time not have been among those contributing to the 2 million *liang*, which may probably also be true for Yunnan and Guizhou, because those two provinces were quite poor and barely able to supply so much money to the war chest, especially after the province of Yunnan had had to organize the Myanmar campaigns a short time ago.

At least for part of the other sums sent to the war chest in Sichuan the origin can be known, yet without that further details about the concrete sums can be learned about. The 4 million *liang*, for example, dispatched on order of QL 39/9/27 (Oct 31, 1774) were to be provided from among the provinces of Henan, Shanxi, 'Hedong' (southern Jiangsu), Anhui, Jiangxi, Hubei, Hunan, and Guangdong. The money from Hunan left the provincial treasury on 10/13 (Nov 16, 1774), that of Hubei on 10/15 (Nov 18, 1774), and the money from Henan on 10/20 (Nov 23, 1774). The last sum had not yet reached the treasury of the province of Sichuan one month later (11/21, i.e. Dec 23, 1774).[93]

In QL 39/5 (Jun 1774) the emperor ordered that 5 million *liang* be provided for the war, and the Ministry of Revenue had the task to select provinces "where money is left" (the part of the tax revenue allowed to be retained in the province, *liucun*), and to organize the transfer of those sums to Sichuan.[94]

In the last year of the war the respective imperial edicts ordered in three cases that the money had to come from the treasury of the Ministry of Revenue, but with the request that the Ministry should "follow the precedents of the years before" and assign the provincial governors to select trustworthy personnel, who were to take care of the transport. Although in this phase of the war the largest part of the war funds was provided by the central government, the provinces had to take over the

---

92 *Gaozong shilu*, juan 896, fol. 30a–30b (QL 36/11/xinhai); juan 918, fol. 8a (QL 37/10/yichou).

93 *Jinchuan dang* 39/III/00339 (QL 39/9/29); 39/IV/00217–218 (QL 39/12/17), 00231 (QL 39/12/19).

94 *Gaozong shilu*, juan 958, fol. 10a (QL 39/5/jiwei).

cost for the transport of the money. Provinces that had hitherto not been involved, like Zhili, also contributed smaller amounts towards the cost for the transport of money, the rations for the troops passing through and the staff of the courier service.

Concerning the other financial transactions, it is stated in many cases that all of the money was disbursed from the treasury of the Ministry of Revenue. This was the case especially in the last year of the war, when it is said in the documents that the money came "from within the treasury of the ministry" or that the funds were "money of the treasury of the Ministry". The same is true for the monetary transfers on QL 37/7/5 (Aug 3, 1772) and QL 38/5/13 (Jul 2, 1773), definitely made directly from the treasury of the Ministry of Revenue. The 3 million *liang* disbursed by the imperial order dated QL 37/2/27 (Mar 30, 1772)—which is only the second largest sum especially earmarked for the troops in Jinchuan—were surely also supplied by the treasury of the Ministry of Revenue. A document relating to this is quite long and contains the emperor's reflection about the actual state finances.[95]

While the treasury of the Ministry of Revenue at the beginning of the Qianlong reign only held 33 to 34 million *liang*, as the emperor writes, it now held 78 million *liang*, only less than one million of which had to be sent (back) to the provinces annually for various purposes, so that during the past ten years or so it had been possible to amass more than ten million *liang*. While before 1755 the Imperial Household Department (*neiwufu*) had be to supported by the Ministry of Revenue with money, expenditure now was somewhat reduced with the result that the Imperial Household Department was able to transfer money to the Ministry. Thus there would absolutely be no need to issue a 'call for contributions' by rich entrepreneurs.

It is not clear if the Imperial Household Department did in the end transfer money to the Ministry of Revenue, but as this would have been a matter of great importance, one could expect to find another imperial decree about this issue. The generally good condition of the state treasury is also referred to in a document dated QL 37/1/15 (Feb 18, 1772) where it is said that "the treasury of the Ministry of Revenue holds 80 million *liang*, and each time there is too much money in the treasury We [the emperor] think that it is a pity". The government should make use of the money which the empire produced. "This large sum [i.e., the 80 million *liang*] is

[95] *Gaozong shilu*, juan 920, fol. 23a–25b (QL 37/11/guimao).

more than one and half as much as We had stored in the treasury at the beginning of the Qianlong reign."[96] Another document dated QL 38/5/9 (Jun 28, 1773) says that there were still more than 70 million *liang* in the treasury of the Ministry.[97] A similar sentence crops up a few months later: "the treasure chamber of the Department is full to the brim",[98] and more than one year later, "the treasuries of the state are full".[99]

In his book about the Imperial Household Department Preston M. Torbert describes its relationship to the Ministry of Revenue during the Qianlong reign as follows: In the first twenty years or so of the Qianlong reign the Department largely obtained money from the Ministry, while from the 1760s on funds were transferred from the Household Department to the Ministry.[100] It can therefore be assumed that the Ministry of Revenue in most cases nominally paid for the cost of the second Jinchuan war, while part of the money in fact came from the funds of the Imperial Household Department. How much it was, however, we do not know, due to the lack of figures.

A great part of the income of the war chest was of course not the monetary transfers coming from the Ministry of Revenue or external provinces. The annual tax income of the land-tax of the province of Sichuan itself was about only 878,000 *liang*[101]—nearly absurd low, as Dai Yingcong says[102]—and at least part of that sum could therefore stay in Sichuan in order to finance the war, instead of being remitted to the central government's treasury. Miscellaneous taxes and the tea tax in Sichuan were also able to contribute at least a quarter of a million *liang* annually.[103] Madeleine Zelin has investigated data from the *Da-Qing huidian* "Collected statutes of the Great Qing" which show that 70 per cent of the tax income of the province of Sichuan stayed in the province and were not sent to the

[96] *Gaozong shilu*, juan 900, fol. 31b–32b (QL 37/1/xinhai).

[97] *Gaozong shilu*, juan 934, fol. 9b–10b (QL 38/5/dingmao).

[98] *Gaozong shilu*, juan 941, fol. 20b–21a (QL 38/8/dingwei).

[99] *Gaozong shilu*, juan 974, fol. 14a–16a (QL 40/1/yimao).

[100] Preston M. Torbert, *The Ch'ing Imperial Household Department: A Study of its Organization and Principal Functions, 1662–1796* (Cambridge: Harvard University Press, 1977), pp. 127–129.

[101] Wang, *Estimate of Land-Tax Collection*, Table 27.

[102] Dai Yingcong, *The Sichuan Frontier and Tibet: Imperial Strategy in the Early Qing* (Seattle: University of Washington Press, 2009), p. 229.

[103] Compare *Huangchao wenxian tongkao*, juan 40, *Guoyong*, juan 2, pp. 5525–5228.

central government. This is partly due to the relatively low land-tax level of that province.[104]

To make things even more complicated, documents contradict each other: While the imperial order dated QL 40/9/29 (Oct 23, 1775) explicitly says that the 3 million *liang* to be transferred will come from the treasury of the Ministry of Revenue,[105] another document shows that only part of the money was procured by the Ministry, and the rest was paid by some provinces.[106] The same is valid for the 5 million *liang* dispatched according to the imperial order dated QL 40/1/22 (Feb 21, 1775).[107] In the end it is clear from the standpoint of logistics that it was more reasonable to transport the money directly from any southern province to its destination in Sichuan instead of via the Ministry, because the money which the Ministry disposed of originated in the provinces, as they delivered their tax income to the central government. From this situation it can be learnt that the emperor avoided to debit the purse of the central government, although it was bulging—as could be seen—and there was "no need to be stingy", as the Qianlong emperor frequently repeated.

The data discussed clearly show that the cost was quite evenly distributed among the provinces and other units of administration, like the salt administration zones. Although not all data for the 'various provinces' are recorded, the sums for the provincial data that are available are all more or less of the same size. Apparently the Ministry of Revenue has paid about two thirds of the total cost, but because we have learnt that even a considerable part of items listed as "paid from the treasury of the Ministry" were in fact paid by the provinces, it must be assumed that probably another third of the large 'piece' of the cake was part of the provincial contributions. Its exact size cannot be determined from the data in the edicts and memorials and surely also depends on the period of time: In the first third of the war there were several financial transactions totally paid by the provinces; during QL 38 (1773) the war was financed by the contributions of rich salt merchants (see below), and in the last third of the war the money came from the budget of the Ministry and from the provinces. Probably it is just the two data from the year QL 40 (1775) that can hint at how large the contribution of the Ministry was, compared with

---

[104] Madeleine Zelin, *The Magistrate's Tael: Rationalizing Fiscal Reform in Eighteenth-Century China* (Berkeley: University of California Press, 1984), pp. 27–28.

[105] *Gaozong shilu*, juan 991, fol. 30a (QL 40/9/jiaxu).

[106] *Junjichu lufu zouzhe*, reel 35, no. 792.

[107] The 3.37 million *liang* were paid by Jiangxi, Guangdong and Gansu.

that of the provinces: Of 5 million *liang* (dispatched on QL 40/1/22, i.e. Feb 21, 1775) the Ministry paid 1.63 million, that is 32 per cent, and of 3 million *liang* (dispatched on QL 40/9/29, i.e. Oct 23, 1775) the Ministry paid 1.2 million, or 40 per cent.

The above considerations are based on the data provided by imperial edicts included in the *Qingshilu* and the *Pingding Liang Jinchuan fanglüe*. Yet the account book of the *Jinchuan junxu li'an* renders figures different from those, sometimes with considerable differences. Assuming the correctness of the official accounting figures in the *Jinchuan junxu li'an* it can be seen that official collections of documents, as the imperial edicts in the *Qingshilu*, are not really a reliable source when it comes to figures of money or personnel. The reason for this is that the imperial edicts are often simple commands to the government agencies to take action in a certain direction. In many cases it was the Ministry of Revenue rather than the emperor or the State Council (*junjichu*) who decided about what action to take in a given situation. Yet the imperial edicts as a documentary source have a decisive advantage: While the amounts mentioned in the *Jinchuan junxu li'an* are absolute sums reflecting the final situation of the finances in a kind of balance sheet, the imperial edicts are dated and therefore provide an insight into the chronological sequence of the particular monetary transfers.

Although some sums are very similar, like those for Guangxi, Shandong, Shanxi, Hubei, and the Zhejiang salt administration, the provinces of Hunan, Henan, Jiangsu, Anhui and the Liang-Huai salt administration in fact delivered much higher amounts than is reflected in the imperial edicts recorded in the *Qingshilu*. Hunan delivered 3.1 million *liang*, and not only 1.8 million; Henan 3.6 million compared to an ordered sum of 1.4 million; Jiangsu paid 3.81 million and not only 1.51 million; Anhui provided a share of 2.3 million rather than 0.9 million; and the Liang-Huai salt administration zone, instructed to send 0.6 million, dispatched in fact 3.4 million *liang* to Sichuan. Concerning the province of Shaanxi, there are not even documents of this kind available in the *Qingshilu*. On the other hand the sum of money provided by the province of Guangdong seems to be much higher in the *Qingshilu* edicts (2.885 million *liang*) than the province in fact supplied (1.8 million). The reasons for the sums being higher in the *Jinchuan junxu li'an* account book could be either that edicts concerning the transfer of sums from the provinces of Hunan, Henan, Jiangsu and Anhui were not recorded in the *Qingshilu* collection (which is, still, a selection of revised and prepared documents), or that the monetary transfers as reflected in the imperial edicts were carried out on an extraordinary basis, while other transfers were made through a different method, for example, by routine transfer via the Ministry of Revenue or directly from province to province, which did not require any special edicts by the emperor. The autonomy of the Ministry of Revenue in financial matters can permanently be observed during the war, when problems occurred for the solution of which no case of precedence could provide a guideline. All decisions over any expenditure would have to be assessed by the Ministry of Revenue and were approved by the emperor only years later, when the *Junxu zeli*, the rules for military expenditure, was compiled A last reason for discrepancies in the figures might be that it is sometimes not clear whether a sum of money came from the provincial treasury (for example, Guangdong or Zhejiang), or from salt administration units whose main seat was located in that province. Documents might simply cite the name of the province without making clear that the respective sum was not provided by the provincial treasury, but came from the treasury of the salt administration commission or the circuit administration, which was not answerable to the provincial governors but to the censorate. Thomas A. Metzger has shown that the salt administration zones, overlapping with the territorial jurisdiction of the provinces and prefectures, were far from having a clearly regulated administrative structure, and were instead rather horizontally incorporated into the local government structure without being subordinated to a provincial governor.[108]

As a result of these investigations it can be said that the Ministry of Revenue paid almost half of the funds coming from outside Sichuan, while the difference to a sum of about 57 million *liang* was contributed by various provinces. The province of Sichuan itself contributed only 2.5 million *liang* to the war chest, due to its low level of land-tax income. This appropriate distribution of the financial burden shows that the Qing government was well aware of the differences in the financial strength of the several parts of the empire and therefore did its best to impose dues suited to the possibilities of each contributor. More than 3 million *liang* were also contributed by different agencies of the salt administration in the lower Yangtse region (Zhejiang, Liang-Huai) which transferred their income from the state-owned salt monopoly to Sichuan. All provinces dispatching financial aid to Sichuan disbursed these sums from their tax

[108] Thomas A. Metzger, *The Internal Organisation of the Ch'ing Bureaucracy: Legal, Normative, and Communication Aspects* (Cambridge: Harvard University Press, 1973), pp. 223–224.

income. In some cases it is known that the salt *gabelle* constituted a great part of the province's revenue that could be used for purposes of that kind. It is furthermore possible that part of the cost taken charge of by the Ministry of Revenue was disbursed from the treasury of the Imperial Household Department, but there is no accessible document so far to prove such a transfer.

The Financial Capabilities of the Provinces and the Empire

Two conditions facilitated the emperor's decision to support the generals in Sichuan with large sums coming directly from the state treasury in the centre as well as the provinces. Firstly, decades of prosperity had filled the treasury of the Ministry of Revenue and the Imperial Household Department to such a degree that the emperor several times stressed that there was no need to be too economical, on the contrary: to hold on to the money would mean to deprive the people of its abilities and expectations—in other words, once the taxes were paid, the money should be used.[109] Secondly, the war against the few thousands of rebellious tribes in the mountains "took too long to give up",[110] and therefore it was especially in the last year of the campaign that the emperor no longer disputed the cost and willingly ordered dispatching money in ever larger sums. For a long period of the war, 5 million *liang* had become a kind of standard sum to be transferred to Sichuan. The emperor even ceased to decide in person what source the money should be disbursed from, instead he simply ordered the Ministry of Revenue to think about that problem. From the documents in the year 1775 we gain the impression that orders to transfer money to Sichuan had become a kind of routine.

A sum of 25 million *liang* found its way from throughout the provinces directly to the war chest in Chengdu, including taxes from the salt administration zones. This money had its origin mainly in the land tax of the particular provinces, a part of which could stay always in the province of origin (the so-called *liucun* share) and had not to be delivered to Beijing. Yet the sums transferred to Sichuan were used for the war and thus could not be used for other purposes, e.g. for local infrastructure. Each year over a period of almost five years the provinces therefore had to go without 5 million *liang*, which means almost half a million *liang* per year and province. A comparison of the tax volume, the amount of retained

109 *Gaozong shilu*, juan 1141, fol. 21b–24b (QL 46/9/dingmao).
110 *Jinchuan dang* 37/III/00144 (QL 37/11/1).

taxes and the contribution which each of the provinces had to make to the war is highly interesting.[111] There is a more or less clear correlation between the total tax volume and the retained taxes, expressed by the fact that the higher the total income through land taxes, the higher the amount of retained taxes. The taxes retained in the particular provinces vary between 11 (Shanxi) and 70 per cent (Sichuan) of the land tax revenue (or rather: the land tax rate), but mostly range around 20 per cent, which means that most provinces were allowed to keep about 20 per cent of their land tax income. Nevertheless, there is no coherent interrelation between the land tax or the retained tax volumes and the contributions the provinces had to make to the war chest in Sichuan. The criterion how to squeeze money out of the provinces was apparently not their regular financial situation. While some 'rich' provinces like Zhejiang paid almost nothing (if not considering the tax income from the salt zone), poor provinces like Hunan and Shaanxi paid proportionally much more. The most eye-catching circumstance is that provinces retaining quite a low amount of land taxes in their provincial treasuries (Guangxi, Guangdong, Hunan) had to contribute money to the second Jinchuan war which amounted to more than double, in one case even almost four times (Hunan) the amount which the respective provinces were allowed to retain from their land-tax revenue. Even the province of Shaanxi which dispatched a huge number of troops to Jinchuan and therefore had to carry the burden of organizing their march to Sichuan, had to pay a huge sum compared to its tax income. The province of Shaanxi furthermore had to organize the courier service and had at the same time the problem of the nearly unmanageable plank road viaducts (*zhandao*) in the Qinling Range.

The annual contribution of the province of Hunan was even almost as high as its annual land tax rate (0.62 million for each year of the war, out of 0.64 million *liang*). This goes to show that the provincial contributions to the war chest cannot possibly have come from the retained taxes but must have been paid from the total tax income. This would have the consequence that in the next year there would be less money or none at all sent to the treasury of the Ministry of Revenue which means that during the war the tax income of the state was far lower than normal. Other provinces were still able to hand over their taxes regularly, like Zhejiang, Shandong, Anhui or Shanxi, to say nothing of Zhili and Fujian, which did

[111] Figures about the tax volumes and retained taxes can be found in Zelin, *Magistrate's Tael*, p. 28, based on the tax quota fixed in 1685.

not have to pay any contributions to the war chest at all. Zhili at least had, like Shaanxi and Shanxi, the duty to organize the march of large contingents of Banner troops to the front and back and to organize the courier service. What counts even more is that some of the provinces paying the highest percentage of their income for the war effort also dispatched large contingents of troops to the front: Shaanxi, Hunan, and Hubei. The province of Guizhou with its extremely low land tax income did not have to provide any financial contributions, just like the province of Yunnan. Yet both sent out many thousands of troops, for which the baggage pay had to be founded. The historical background of that region must not be forgotten: only a short time ago the imperial troops had tried to invade northern Myanmar from there, and for the organization of that campaign, the province of Yunnan had been responsible, although it had not contributed financially to the 13 million *liang* which the Myanmar campaigns had cost.[112]

Another case has to be mentioned: the province of Sichuan itself contributed more than 2.5 million *liang* to the war, but the annually retained tax of Sichuan was just about one tenth of the money to be paid to the war chest. This could only be done if either the central government waived taxes to a large degree, or if the province of Sichuan paid off those debts to the central government over a long period of time, probably by raising some minor additional taxes which normally served to finance the bureaucracy of the local government, for example the *jintie* surcharge which served to finance the subsidies for governmental labourers. A whole series of tax abatements for various districts in the province of Sichuan was therefore implemented.

The first tax abatement during the war took place in QL 37/1 (Feb 1772). A year earlier there had already been a postponement of tax collection alternating among several districts and it was ordered that the miscellaneous and grain taxes of the Sichuan residents and the tributes of the 'barbarians' would uniformly be collected only one year later. Yet since six months before that people of Sichuan had been recruited for logistics service and the emperor therefore ordered the governor-general to consider a second general extension of the tax collection after the end of the war (which was, at that point of time, still not foreseeable).[113] Only half a year later it seemed better to immediately apply a policy of tax

[112] Figure according to Lai, *Qianlong zhongyao zhanzheng*, p. 426.
[113] *Gaozong shilu*, juan 900, fol. 3a–3b (QL 37/1/wuxu).

respite because the intensification of the war made it necessary to alleviate the load of the population of Sichuan relatively soon. A'rtai, at that time acting governor-general of Sichuan, was therefore ordered to arrange a procedure for tax abatements and tax extensions to ease the burden on the population.[114] Again half a year later, just before the first occupation of Lesser Jinchuan, the emperor ordered to postpone the tax collection of fifty prefectures and districts where troops had passed through, from 1773 until the year 1774. Ninety further prefectures and districts which did not have to sustain the passage of troops but whose inhabitants had to provide labourers, were allowed to pay their taxes from 1772 two years later, in 1774. Inhabitants of territories beyond the border through which troops had passed, were likewise allowed to pay their 'taxes for barbarians' and tributes (grain, horses) one year later than normally.[115] Another six months later, after the catastrophe of Mugom, when large contingents of elite troops from the Capital and the northeast were dispatched to the front, the emperor again postponed the tax collection of those districts, through which troops had passed, until the year 1774. And now, for the first time during the war, the emperor even included other provinces into his tax postponement programme. Affected districts in Shaanxi and Gansu, which had already been granted the favour to pay their taxes later, were now allowed to delay the payment of taxes by a full year, if 50 per cent of the collected taxes had been extended; 80 per cent were to be postponed for those districts which had before been allowed to postpone 40 per cent; and the districts which hitherto had been allowed to pay 30 per cent of their taxes later, were now allowed to be spared 60 per cent for the current year because the population had to provide services to the troops and to lend out carts and beasts of burden. The affected districts in Zhili and Henan were granted a tax extension of 50 per cent.[116] A few months later eleven more districts on the march route were granted a tax extension of 60 per cent.[117] Later on, the logical consequence of the benevolent methods hitherto applied was to convert the tax extension into a tax abatement. Therefore, the emperor ordered the governor-general of Sichuan to check what number of labourers those districts had

---

[114] *Gaozong shilu*, juan 910, fol. 8b–9a (QL 37/6/renshen). *Pingding Liang Jinchuan fanglüe*, juan 30, fol. 15b–16a (QL 37/6/renshen).
[115] *Gaozong shilu*, juan 924, fol. 3b–5b (QL 38/1/renchen).
[116] *Gaozong shilu*, juan 937, fol. 25b–27a (QL 38/6/renzi). *Pingding Liang Jinchuan fanglüe*, juan 63, fol. 20b–21a (QL 38/6/xinhai).
[117] *Jinchuan dang* 38/III/00459 (QL 38/9/3).

provided that had also been granted a tax extension until that date. Those having provided a large number of labour conscripts were to be granted a total tax abatement instead of the tax extensions, and those with a smaller number of labour conscripts were allowed an abatement of 50 per cent on their extended taxes, or a smaller percentage. The rest of the money had to be paid without any concessions.[118] Fulehun and Wenshou thereupon suggested a complex procedure for ninety districts,[119] in which the payment of taxes for three years was, differing from district to district, either extended or abated. A short time later, a similar plan was developed for a number of other locations in Sichuan, including the territories of the native kings who paid their tributes to the Qing court.[120] For the year 1775, the taxes in many districts should likewise be alleviated, yet there were also some districts for which the abatements were transformed into a less preferential treatment.[121] From the last document, as well as from the one issued on QL 38/1/2 (Jan 24, 1773) it is possible to calculate how much money the treasury of the province and the Ministry of Revenue lost. With the help of the land tax rate cited in the *Sichuan tongzhi*, the local gazetteer of the province, the particular amounts can be found out.[122] For the group of fifty prefectures and districts mentioned in the edict dated QL 38/1 (Jan 1773), 252,531 *liang* were missing for one year of tax abatements. For the second group of districts, which were less affected by troops crossing the territory, an amount of 401,629 *liang* was missing annually, when waiving all land taxes.[123] In the *Sichuan tongzhi* chapter about tax abatements (*juanzhen*) the edicts are reproduced which are also contained in the *Qingshilu* collection. No further information, like the total sum of

118 *Gaozong shilu*, juan 967, fol. 41b–42b (QL 39/9/yihai). *Jinchuan dang* 39/III/00331 (QL 39/9/25); 39/IV/00005 (QL 39/11/6).

119 *Gaozong shilu*, juan 970, fol. 8b–9b (QL 39/11/bingchen). *Pingding Liang Jinchuan fanglüe*, juan 109, fol. 17b–18a (QL 39/11/bingchen). *Jinchuan dang* 39/IV/00089 (QL 39/11/7).

120 *Gaozong shilu*, juan 973, fol. 1b–2b (QL 39/12/yiwei). *Pingding Liang Jinchuan fanglüe*, juan 111, fol. 13b–14b (QL 39/12/bingwu). *Jinchuan dang* 39/IV/00207 (QL 39/12/16).

121 *Gaozong shilu*, juan 1005, fol. 6b–8b (QL 41/3/xinmao). *Pingding Liang Jinchuan fanglüe*, juan 134, fol. 1a–3a (QL 41/3/jiayin).

122 A local gazetteer (*difangzhi*) is a chronicle arranged in a specific pattern of chapters on geography, statecraft and administration (astronomy, food and commerce, education, military, offices, examination and promotion, etc.), culture (writings, temples, ancient sites, local products, omina and portents) and biographies of eminent persons (officials, writers, thinkers, 'worthies', eminent women, monks, anachorets, and so on).

123 (Jiaqing) *Sichuan tongzhi* (Chengdu: Ba-Shu shushe, 1984), juan 63, fol. 3a–38b. Data from the year QL 18 (1753).

tax deficits during the war, can be found.[124] Yet the descriptions in the imperial edicts permit to guess that roughly 1.5 million *liang* of taxes were waived because of the second Jinchuan war. To this sum the general tax extensions (*pujuan*) of the years QL 35 (1770) and QL 42 (1777) must be added. In 1777 it was decided to totally waive the levy of Sichuan province of the year *wuxu* (QL 43, i.e. 1778). In 1770 the tax extensions or tax abatements had cost the state treasury 668,400 *liang* in Sichuan, while the tax abatement issued in 1777 was effected in three steps for different regions of Sichuan during the years *xinhai* (1791), *renzi* (1792) and *guichou* (1793), and cost the state treasury 692,300 *liang*.[125]

It is an interesting question if war expenditure and all those tax abatements affected the state revenue at all. The Mongol Faššan (Chinese rendering Fashishan) has included a list of the annual state reserves between the years YZ 1 (1723) and QL 39 (1774) in his essay collection *Taolu zalu* "Miscellaneous notes from the Potter's Hut".[126] Unfortunately these end just in the middle of the second Jinchuan campaign, so that all that might be said about are speculations. Nevertheless it can be seen that the state reserves in the Ministry of Revenue drastically declined in the mid-Yongzheng period (late 1720s), then caught a level of about 30 *million* liang that was again rising visibly from the mid-Qianlong reign (1760s) on.

The year QL 38 (1773) seems to have seen a maximum of almost 80 million *liang* of reserve. If warfare had an impact on the state reserves at all, it can be seen in the years QL 13–14 (1748–1749), when the first Jinchuan war was fought, and in QL 24 (1759), during the war against the Khojas in Kašgar and Yarkant. The period of the drastic drop in YZ 9–YZ 11 (1731–1733) only experienced the campaigns against the Miao insurgents in Guizhou, while the reserves had been raising steeply during the Yongzheng emperor's Dzungar campaigns. The drop at the end of the graph might point at a shortage in funds during the second Jinchuan campaign, but not one that critically affected the state reserves—at least not yet.

Contributions for the Second Jinchuan Campaign

The general accounting book in the *Jinchuan junxu li'an* lists a sum of 5.6 million liang of contributions, 2 of which came from the salt merchants

124 (Jiaqing) *Sichuan tongzhi*, juan 73, fol. 34a–40b.

125 (Jiaqing) *Sichuan tongzhi*, juan 73, fol. 32b–34a. *Gaozong shilu*, juan 1025, fol. 43b–44a (QL 42/1/yiwei).

126 Faššan, *Taolu zalu*, juan 1, 24–26.

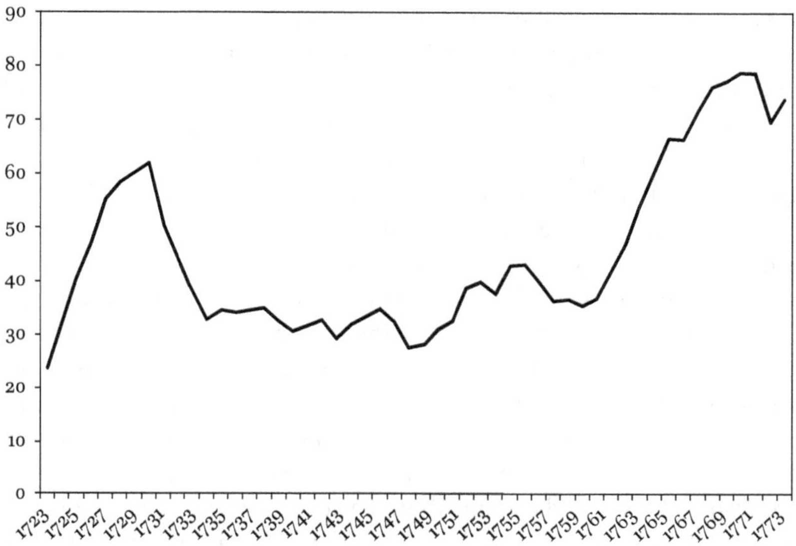

Source: Faššan, *Taolu zalu*, juan 1, pp. 24–26. Time frame YZ 1–QL 39.

Diagram 1. Balance sheet of the Ministry of Revenue 1723–1774 (figures in million *liang*).

in Liang-Huai, 900.000 from those in Changlu, 1 million from Zhejiang, 1.1 million *liang* from Shanxi, 200,000 from Guangdong and 400,000 from Guangxi. The texts speak of 4 million *liang* to be delivered by the merchants of Liang-Huai, yet only 2 million were paid until the end of the contribution campaign. Most of these sums can be confirmed by imperial edicts included in the *Qingshilu*. The salt wholesalers (*gangshang*) from Liang-Huai contributed 4 million *liang*.[127] As already heard earlier, it often happened that the merchants associations were not able to pay such a huge amount at one go and had therefore to borrow the money from the state treasury that they were to pay to the same, in other words, the money was paid in instalments over a period of four years. Yet there was a dispute between salt supervisor (*yanzhengshi*) Li Zhiying and the emperor about the time of the 'reward' to be handed over to the donators. While Li was of the opinion that they first had to pay the whole sum before they could be ranked for a reward and were given a patent for the brevet title for a state office, the emperor said that this would take too

[127] *Gaozong shilu*, juan 941, fol. 21b–22a (QL 38/8/dingwei).

long and ordered compiling the list of merits immediately.[128] In Zhejiang, governor Sanbao submitted a palace memorial reporting that the local merchants 'desired' contributing 1 million *liang*. The money donated with "burning enthusiasm" was to be paid within five years' time.[129] The salt supervisor of Changlu, Xining, submitted a palace memorial in which he reported the donation of 600,000 *liang* by the salt merchants of Changlu and 300,000 *liang* by the merchants of Shandong.[130] Members of the gentry of Shanxi asked governor Bayansan to be allowed to donate 1.1 million *liang* as "transport capital". These bankers promised to dispatch representatives who personally took over the transport. Their "double effort", not only to contribute money, but also to take over organization and cost of the transport, was accordingly to be rewarded with "extraordinary positions" in the list of promotions.[131] Salt merchants of the Liang-Guang region (Guangxi, Guangdong) submitted the request to governor-general Li Shiyao to be allowed to contribute 200,000 *liang*. Overseas merchants also contributed 20,000 *liang*.[132] As could be seen with the Mongol allies, it was very common that even native rulers offered 'tributes' in the shape of contributions. The native queen of Somo and her son contributed five hundred cattle, one thousand baskets of wine, and five hundred loads of *tsampa*. The native prince was rewarded with a promotion from second-class pacification commissioner (*anfu*) to first-class pacification commissioner (*xuanwei*).[133]

Money was in fact the original currency donated by the contributors, but because the main reason for contributions was the purchase of rice to feed the army, the example of the Shanxi gentry clearly demonstrates that it was not money that should be transported to Jinchuan, but commodities. For the government side it was important and of great help that the donators did not only contribute money or rice, but also undertook the task of transporting the rice to the doors of the military camps, thus virtually taking over functions normally carried out by the government.

---

128 *Gaozong shilu*, juan 943, fol. 22b–23b (QL 38/9/dingchou).

129 *Gaozong shilu*, juan 943, fol. 31b (QL 38/9/gengchen).

130 *Gaozong shilu*, juan 944, fol. 15a–15b (QL 38/10/yichou). *Jinchuan dang* 38/IV/00035 (QL 38/10/4).

131 *Gaozong shilu*, juan 944, fol. 18b–19b (QL 38/10/gengyin); juan 956, fol. 8b–9a (QL 39/4/wuzi). *Jinchuan dang* 38/IV/00041 (QL 38/10/6).

132 *Gaozong shilu*, juan 948, fol. 27b–28a (QL 38/12/wuxu); juan 949, fol. 18b (QL 38/12/dingwei).

133 *Gaozong shilu*, juan 996, fol. 6b–7a (QL 40/11/dingchou). *Pingding Liang Jinchuan fanglüe*, juan 128, fol. 3b–4a (QL 40/11/dingchou).

Governor-general Fulehun, by the way, therefore directly asked if the contributions by merchants from other provinces could not be handled by paying money instead of rice to the Ministry which would only be 'converted' into rice when the merchants arrived in Sichuan.[134]

Because of their considerable size the 'donations' by merchant associations are especially mentioned in memorials and had to pass through certain bureaucratic institutions before being processed. The governmental authority responsible for the administration of the salt merchants was a salt supervisor (*yanzhengshi*) who officially received the permission which he forwarded to the emperor. It seemed to be of great importance that the Ministry of Revenue and especially the Contributions Bureau (*juannaju*) created a ranking of each person and the respective sum of money contributed. Nearly all the documents quoted above contain imperial orders to create detailed rankings of all persons that had contributed to the donation, and not simply to take the whole sum and to use it. This sort of often repeated imperial instruction is not just a kind of perfectionism but shows that the *juan* was above all really seen as a kind of tax whose payment had to be registered exactly. Second place took the consideration that the *juan* contribution served to assess a precise merit that had to be rewarded adequately. It was necessary to check how much somebody had paid or promised to pay in order to be able to be listed among the applicants for a reward. Therefore the emperor asked in several instances for detailed reports about the necessarily different financial performance of each member of the contributors' group.

Although the petitions by the merchants on the surface look like voluntary donations of the rich, the state systematically tried exploiting new sources for the financing of its expensive undertakings. The petition of the Shanxi gentry that was handed in to the court as a memorial by governor Bayansan, has a past history. At the end of the summer in 1773 the emperor, when summing up how much money he has already sent to Sichuan, came up with a sum of about 10 million *liang*, enough to continue the war until the beginning of the next year or even until summer 1774. The problem of providing finances is here connected with the problem of logistics. Bottlenecks in the transport of provisions and war material have already been solved by entrusting the transport to entrepreneurs who hired porters and cared for the transport to the doors of the military camps. The entrepreneurs would then charge the state with the cost for

134 *Gaozong shilu*, juan 946, fol. 10b–11a (QL 38/11/xinyou).

material and transport. This way the generals would know how expensive it was to supply the army and would be able to calculate the financial needs for the following year. The most suitable group of merchants or 'rich people' that could undertake such a transport were wealthy families from Shanxi that were obviously known for their capital which could stand the state in good stead. Fulehun and Wenshou were ordered to think about a possibility of making use of the money of the bankers in Shanxi, and the provincial governor Bayansan was told to more or less openly attract investments by offering to reward donators with licences for offices of a higher rank or to put them on top of the list for the assessment.[135] The petition of the Shanxi members of the gentry is therefore not a result of their free will to provide the state with a large sum for a war in the Sichuan hinterland but rather something enforced by the government as a kind of taxation of the rich Shanxi business people. Although there are not many documents left about these contributions, there is a hint in one document that also members of the gentry (here called *shimin*) in other provinces were able to disburse money as contributions with which they bought rice and had it transported to assigned logistics stations.[136] The financial performance of each contributor was assessed according to the transport prices—more closely situated logistics stations were of course cheaper than those far in the high mountains. The war logistics bureau had then to find out the best transport routes for those contributors.

The emperor, answering the "requests" of the donators, largely explained in his edicts that the state actually did not need any help by loyal and faithful subjects but because there were precedents asking for an equal treatment of similar cases, and because the donators "beseeched Us so enthusiastically and urgently" the emperor finally accepted the donation. The emperor's concern was in fact to use as little 'outside' money as possible and instead to exploit all sources of the state founded on a regular basis. Therefore *juanshu* contributions should only be used sparingly and when the government was in a state of distress. A good example that the emperor was also able to decline 'petitions' for a contribution campaign

---

[135] *Gaozong shilu*, juan 938, fol. 30b–32a (QL 38/7/yichou). More information about the Shanxi bankers can be found in Yang Lien-sheng, *Money and Credit in China: A Short History* (Cambridge, Mass.: Harvard University Press, 1952), or Liu Kwang-Ching, "Chinese Merchant Guilds: An Historical Inquiry", in *Pacific Historical Review*, 57/1 (1988), pp. 1–23. See also Andrea Lee McElderry, *Shanghai's Old Style Banks (Ch'ien Chuang), 1800–1935: A Traditional Institution in a Changing Society* (Ann Arbor: Center for Chinese Studies, University of Michigan: 1976).

[136] *Gaozong shilu*, juan 943, fol. 6b–7b (QL 38/9/guiyou).

is a memorial presented in QL 41/10 (Nov 1776), when the Jinchuan war was already over. There was no further need to rely on contributions by rich merchants or state officials because the treasury of the Ministry of Revenue held more than 60 million *liang*, and there were millions of *liang* left over in the war chest. The responsible official, salt supervisor Yiling'a, was reprimanded for his precipitate 'fund raising'.[137] Yet even during the war, the emperor tried everything to avoid contribution campaigns. In QL 37/11 (Dec 1772) governor-general Wenshou suggested proclaiming a contribution campaign in Sichuan and was reprimanded by the emperor.[138] At that time the conquest of Lesser Jinchuan made good headway and the war soon seemed to come to an end. The second reason was that the transport of grain to the camps was more and more entrusted to private entrepreneurs (*shangyun*), and it did not seem appropriate to the emperor to establish two different types of transport by private people, one of which would be the contributors taking their donations in the form of rice directly to the war theatre. But the situation was totally different after the catastrophe of Mugom in the summer of 1773, and therefore the emperor allowed to 'open' contribution rounds in the province of Sichuan.[139] The emperor's scepticism about the efficiency of such a direct delivery of rice from contributors to the border was absolutely justified: it was soon reported that contributors did not deliver the rice or supplied rice of inferior quality.[140] The worst case of contributions in Sichuan was when A'rtai and his entourage had extorted money from merchants or landlords under the pretext of contributions (hence termed *lejuan* 'extorted contributions') and with the purpose to finance their luxury spending.[141]

The official imperial edicts only record the larger contributions from merchants or merchant associations. Evidence for less important contributions is rarely produced. Yet two documents can be found mentioning contributions made during the years of the second Jinchuan war: An unknown amount of rice had been contributed in Sichuan from an unknown source. Of course, this rice could directly brought to the camps and was not to be converted into silver via the Ministry of Revenue.[142]

137 *Gaozong shilu*, juan 1018, fol. 21a–22b (QL 41/10/guichou).
138 *Jinchuan dang* 37/IV/00183 (QL 37/11/12).
139 *Jinchuan dang* 38/III/00547 (QL 38/9/27).
140 *Jinchuan dang* 38/IV/00047–48 (QL 38/10/7).
141 This incident, reported in *Jinchuan dang* 38/IV/00271 (QL 38/12/11), must have taken place before the war. It was only detected in connection with A'rtai's bad performance during the war and his subsequent dismissal.
142 *Jinchuan dang* 38/IV/00047 (QL 38/10/7).

Concerning contributions in kind submitted in Sichuan, the government just decided to transfer contributions actually submitted to academic institutions to the war chest to feed the army.[143] In the district of Yongchuan (see Map 3, p. 195) 10,000 *liang* of contributions were collected for repair work of the city moat. As so many people had been recruited to serve as military labourers they could not be recruited for the repair work. Half of the moat had been repaired, and the rest of the work thus had to be postponed until after the war.[144] A group of members of the gentry from the district of Gongxian had contributed 8,390 *liang* to repair the city wall. This sum should instead be allocated to the provincial treasury, probably with the intention to use it for other purposes, e.g. to finance the war.[145] At least part of the money transferred to Sichuan also came from regular smaller contributions in different provinces, but we do not exactly know what they amounted to and in what areas the contributions had been presented. The war logistics bureau directed the respective rice contributions to fixed destinations, and the rice had to be taken there within a certain time, which could be checked by a kind of delivery note (*zhaopiao*). The contributors were to be rewarded according to the amount of rice they had donated and the cost for the transport, which could be very different, depending what camp the grain had to be shipped to.[146]

Lai Fushun has detected two further documents reporting contributions.[147] One is a small donation of 3,000 rider jackets by a textile manufacturer in Suzhou. Another source speaks of only 2,000 rider jackets and war tunics, fabricated like the standard cotton coats of the army which were to be sent to Sichuan in the following spring.[148] A third source renders the figure of "a contribution of 3,000 manufactured additional coats".[149] The other case are 100,000 *liang* provided by salt merchants in Jiangxi. The contribution by merchants from Hangzhou, Jiaxing, Shaoxing and Songjiang is cited by Lai with a very late date, QL 39/2/26 (Apr 6, 1774). In two cases the provincial treasury advanced 800,000 *liang* and the circuit treasury 200,000 *liang*, two sums that immediately were to be sent to Sichuan by seven high local officials. The merchants themselves were given five years to pay back this loan before being listed and ranked

143 *Jinchuan dang* 38/III/00060 (QL 38/7/9).
144 *Gaozong shilu*, juan 955, fol. 27a (QL 39/3/renwu).
145 *Gaozong shilu*, juan 996, fol. 39b–40a (QL 40/11/bingxu).
146 *Gaozong shilu*, juan 943, fol. 6b–7b (QL 38/9/guiyou).
147 Lai, *Qianlong zhongyao zhanzheng*, pp. 418–421.
148 *Jinchuan dang* 38/IV/00315 (QL 38/12/16).
149 *Jinchuan junxu li'an*, juan 2, fol. 185b.

by the ministries of Revenue and Personnel. This case is just the opposite of the case of the Liang-Huai merchants mentioned above who were allowed to be ranked in the files for the Ministry even before the sum had been paid back in full to the government.

Including all smaller contributions it seems that the contribution campaign seems to have yielded more than the 5.6 million *liang* mentioned in the general account (see Table 7, page 122).[150] This only includes the larger amounts of contributions that were confirmed and accepted via palace memorials and imperial edicts. Together with all other contributions delivered by private persons or state officials in the province of Sichuan a sum of probably 10 million *liang* was yielded, as can be found out in several edicts, where it is said that "From the Sichuan transport contribution campaign We have furthermore obtained 10 million *liang*, which is more than enough for the whole of next year".[151] With monthly cost of between 1 and 1.5 million *liang* the money collected by the contribution campaign might have sufficed to feed and equip the army for 7 to 10 months. One edict even states that the funds were enough to finance the war "for another year or two".[152] It can in fact be observed that the contribution campaign was conducted at a time when no other funds were transferred to the war chest, neither from the Ministry of Revenue nor from other provinces. During the time of the financial gap between QL 38/5 (Jun–Jul 1773) and 39/5 (Jun–Jul 1774) the war seems to have been financed by the contribution campaign exclusively (compare Table 9, page 128).

Conclusions

Though the progressive standardization of accounting had probably not decreased the cost of warfare, it did reduce the need to resort to highly unpopular ad-hoc measures to raise funds as it had been common during the early Qing period. Arbitrary commandeering of food and transport tools or advance payment of taxes is rarely reported in documents from the high Qing period. The intensified bureaucracy also gave the government a tighter control on the funds flowing from the garrisons, the provinces, the

---

150 *Jinchuan junxu li'an*, juan 2, fol. 177a–177b.

151 *Gaozong shilu*, juan 949, fol. 12b (QL 38/12/jiachen). *Pingding Liang Jinchuan fanglüe*, juan 84, fol. 10a (QL 38/12/jiachen). Similar statement in *Jinchuan dang* 40/I/00007 (QL 40/1/7).

152 *Jinchuan dang* 38/IV/00031 (QL 38/10/4).

salt administration zones or the private purses of merchant associations or individual officials into the war chest. It allowed the central government a better overview over the sums the provinces disposed of so that it was easier to channel the flow of these amounts of money to spots where they could both serve the cause of the army and the image of the emperor as a benevolent patriarch. Corruption could likewise, at least to a certain extent, be limited. A huge and costly enterprise like a war was financed by a complex network of tax transfers between the different treasuries of the provinces, salt administration zones, the Ministry of Revenue, and the Imperial Household; this intricate "fiscal logic" is in many respects comparable to the management of expenditure in the old imperial structure of the Spanish empire in the Americas, rather than to the more modern empires of the British and the French.[153] Very revealing is the growing part that institutionalized contributions played in the finance of warfare. While in earlier ages contributions consisted of a multitude of smaller sums donated by individual officials, the Qianlong reign saw the factual requisition of ever larger sums of money from associations of government merchants (*guanshang*) as a pre-calculated source for war financing.[154] Contrary to repeated assertions that such contributions were only an emergency solution to be used with caution, the traditionally despised merchants were regularly taken advantage of to help finance the dynasty's mostly hailed profession, that of the warrior, and the civilian officials of the local government were fleeced to pay for funds lacking to finance the supplies of the military machine. Both, contributing merchants and civilian officials, were furthermore also instrumentalized to care for the matter of military logistics, as will be seen in the next chapter.

---

[153] Compare Carlos Marichal, *Bankruptcy of Empire: Mexican Silver and the Wars Between Spain, Britain and France, 1780–1810* (Cambridge: Cambridge University Press), p. 29.

[154] The term 'government' or 'official' merchant refers to their function as merchants endowed with a licence for the marketing of certain goods subject to state monopoly, like salt, or for the exclusive trade with foreigners.

CHAPTER THREE

WAR LOGISTICS

> This time plenty of labourers arrived in the camps, and the Mongol tribes in the vicinity of the logistics stations delivered a sufficient number of horses, so that the public cause could be pursued diligently. They have therefore to be graciously rewarded [...] with bolts of cloth and leaves of tea.[1]

As could be seen in the first chapter logistics was the most expensive item on the list of war expenditure. This circumstance wholly corresponds with the great tactical importance of logistics for all campaigns, in bringing the troops into the war zone at the right moment and at the proper place, to have them supplied with food, transport animals, as well as ammunition. The procurement of all this could have been even more expensive if the Qing government had not made extensive use of the civilian bureaucracy to prepare the march of the troops to the war theatre, to provide carts, beasts of burden or porters, to care for grain, 'salted vegetables' and tea leaves, to organize gunpowder and metal for bullets or cannon, and, if consequently carried on, also to raise funds.

Some of the border regions in the west, especially Sichuan and Gansu, were over decades so predominantly occupied with the organization of the supplies for armies that they virtually "lost their civilian governors in the middle of the eighteenth century, making place to a military administration under the governor-general". The civilian provincial governor thus "fell under the shadow of the powerful and pivotal military governor general", and was "outranked in his own capital".[2] In 1764, when the war against the Dzungars was finally completed, Gansu, with its large amount of magazines specifically built to feed the army, had undergone such a change towards militarization that the province no longer required a governor. In Sichuan, too, the Qing dispensed with the civilian post of governor in 1753 and vested the single figure of the governor-general with the authority over the military and civilian realms. The reason for

---

[1] *Pingding Zhungar fanglüe, Zhengbian*, juan 2, fol. 4a–4b (QL 19/3/yimao).

[2] Guy, *Governors and their Provinces*, pp. 209–211.

these decisions was that the role to control, to supply, and to finance the troops who defended or extended the boundaries of the empire "required more authority than the procedures of a civilian government normally allowed".[3]

In this chapter it will be shown that the bureaucratization of warfare during the Qing period went so far that in the late eighteenth century campaigns could be launched quickly with the efficient support of the civilian bureaucracy. With the help of statutes and regulations, and based on the experience of several generations of continuous warfare in the southwestern provinces and the border regions, the logistic organization of campaigns became virtually a matter of routine.

War Logistics in the Seventeenth and Eighteenth Centuries

The experience made during a number of wars that had been fought in the same regions contributed to a smoother organization of the campaigns during the eighteenth century. Over a long period many campaigns were undertaken in Eastern Turkestan as well as in Qinghai and Central Tibet: the wars against Galdan, khan of the Oirats, between 1677 and 1696 and against his nephew and successor Tsewang Rabdan 1715–1717, the campaign of 1730–1732 against the Dzungars, the two final campaigns of 1755 and 1756–1757 against the same enemy, the war against the Khojas in Yarkant and Kašgar in 1758–1759, the campaigns to relieve Uchturpan (Uš) from the hands of the rebel Rehmetulla Beg in 1765, and the two wars against Muslim insurgents in 1781 in Lanzhou and 1784 in Fort Shifeng (Shifengbao), explored the western regions in detail and made it possible that several generations of commanders knew the territory and its challenges throughout. In 1718, 1727 and 1750 campaigns led into Central Tibet, and then from 1788 on again until the pacification of the Gurkhas from Nepal in 1792. Qinghai was conquered in 1721–1724 and incorporated into the Qing empire. In the eastern parts of Tibet, several smaller campaigns in 1700 (Dartsedo), 1745 (Djandui), 1748–1749 (first Jinchuan war) and 1752 (Dzagu) initiated the subjugation of the local kingdoms, ending in the long second Jinchuan war 1771–1776. In southwest China there were many uprisings of the native Miao tribes against increasing Chinese colonization in Hunan, Guizhou and Yunnan, especially in the

[3] Guy, *Governors and their Provinces*, pp. 215–217, 228.

first decade of the eighteenth century, the 1720s and the 1730s. Yet the early uprisings of natives in these provinces in the 1660s as well as the protracted suppression of the Three Feudatories have also to be counted into the logistics pattern of this region. These wars were fought internally, inside provincial territory and therefore not considered as 'wars', meaning an operation against a foreign enemy that encroached upon the rights of the empire, but as constabulary actions to put down a rebellious group of the population as part and parcel of the Green Standards' duties.[4] Because these campaigns were undertaken inside imperial territory the problem of logistics therefore does not seem to have been as serious as for the wars fought far away in Eastern Turkestan or Central Tibet. Some other campaigns were quite extraordinary compared with those fought in these regular war zones and the logistics managers could therefore not rely on precedents. Such wars were the three campaigns to conquer and reconquer Taiwan in 1683 (the Zheng dynasty), 1732 (rebellion of Lin Wuli and Wu Fusheng), 1770 (rebellion of the sectarian leaders Chen Zongbao and Zheng Chun) and 1786 (rebellion of Lin Shuangwen), as well as the Myanmar campaigns of 1766–1769, and the Annam campaign in 1788. At least the experiences made in Taiwan might have resulted in a smoother proceeding in the suppression of Lin Shuangwen's rebellion in 1786. The second Dzungar campaign of the Qianlong emperor, his second Jinchuan campaign and the second war against the Gurkhas from Nepal appeared as excellently organized wars in which problems with the logistics rarely occurred. Yet there are also some examples of logistic failures that were partially due to formerly unknown topographical conditions, as was the case with the Myanmar campaigns and the war against Lin Shuangwen in Taiwan, and partially due to poor organization, as in the first Jinchuan war and the Annam campaign. While there is not much known about the logistic challenges of warfare in the earlier part of the Qing period, Lai Fushun was able to collect a wealth of information on the issue of logistics during the Qianlong reign.

For the preparation of the march towards the war theatre the logistics managers had developed particular statutes. Precedents for the organization of the march to the war theatre and back date at least from the Kangxi reign. It was the task of the local governments to organize all necessary arrangements, which were called *bingchai* 'military duties', to be fulfilled by the local communities. Statutes or precedents were often

[4] Luo, *Luyingbing zhi*, p. 252.

amended or improved, according to experience, and if no precedents for a certain province of region were available, those of other provinces or regions had to be adhered to, with adaptions, should the occasion arise. The existence of such regulations was very important because troops of different types (Capital Banner troops, provincial Banner troops, Green standard troops), and officers of different ranks were entitled to different numbers of horses and took with them varying numbers of manservants and bulks of luggage, as could be seen in Chapter One.

The organization of a march of thousands of troops from their garrisons to the war theatre was a tremendously complicated task because—ideally—the prefects or district magistrates whose territory the troops passed had to take care of the transport of baggage from their own district to the next, the accommodation of the troops, and their food, as well as fodder for the horses. Therefore it had to be exactly pre-calculated at what time what number of troops would arrive, spend the night, and march on. For the transport of baggage and military equipment the local government had to assemble a sufficient number of carts or beasts of burden. In order to feed the troops, sutlers had to be invited, but also money-changers, because the troops were not necessarily paid out copper cash for their daily rations, but often enough silver, and had to change this into cash beforehand. For accommodation, an appropriate location had to be prepared where the troops would erect their tents. When rivers had to be crossed or waterways were chosen to travel on, boats had to be organized.

The troops were expected to cover a distance of about 100 *li* per day (1 *li* corresponding to about 500 metres) on their way to the encounter with the enemy, and they spent the night in logistics stations (*taizhan*). In mountainous regions where only shorter distances could be covered per day, these stations were accordingly to be found closer to each other. There the distance could be as short as 50 or 60 *li*, yet on level ground logistics stations were as far away from each other as 140 or 150 *li*.[5]

The process of organizing the march to the front was normally initiated by an imperial edict issued after deliberation with the State Council (*junjichu*) about which troops to send to the war theatre and what routes to take. This edict was sent to the governors-general, who had to "oversee

[5] *Hubu junxu zeli*, juan 5, fol. 1a. In Europe, infantry troops, accompanied by the baggage train, normally marched about 15 miles a day, according to van Creveld, *Supplying War*, pp. 2, 12.

the projection of Qing power into the countryside",[6] and to the provincial governors in question who were responsible for their part of the route of the march, together with the provincial military commanders (*tidu*). According to the statutes, larger cities were to be avoided in favour of market towns, where it was easier to assemble the required amount of carts, mules and merchants. Scrupulous attention was paid to avoid too much contact between the troops and the local population, presumably in order to prevent them from harassing the latter and to ensure discipline among them, or possibly in order to prohibit the locals from trying to distract the troops from their actual business, the preparation for war. Alternatively, a camp could be set up outside inhabited territory where there was no danger of incidents between civilians and soldiers at all, yet for the preparation of such a place the organization requirements were probably higher. The most suitable choice was, of course, to use one of the courier stations (*yizhan*) for lodging, because they were permanent facilities with the necessary equipment and experienced personnel. In larger (*dazhan*) or 'proper' logistics stations (*zhengzhan*) the soldiers were fed and could change their riding horses, while there were also smaller stations where the troops could spend the night and were given food, but no fresh horses. Smaller (*xiaozhan*) or 'interstitial' stations (*yaozhan*) had the advantage that no horses had to be organized, yet they also posed the danger that exhausted animals dropped or damaged wheels could not be repaired. There were also some 'snack' stations (*jianzhan*) in which only food was provided, but neither horses nor shelter for the night. The speed of the march and the frequency of troop contingents passing a station were decisive factors for the organization and the feasibility of the operation.[7]

Carts and horses organized in one station served to take a batch of troops to the next station. The transport vehicles and animals then returned to base, where they waited for the next batch of troops to be received and accompanied. The intervals between the batches were between three and five days. These intervals as well as the size of the batches were to be fixed in advance, so that the correct amount of carts, mules and horses necessary to shuttle the troops from one station to the next could be arranged. If the intervals were too short, carts and horses did not have enough of a break; if they were too long, cost accrued without benefit because the

[6] Guy, *Governors and their Provinces*, p. 49.

[7] The description of the stations and their organization is based on Lai, *Qianlong zhongyao zhanzheng*, pp. 53–64, who himself relies to a large extent on the *Hubu junxu zeli*.

owners of carts and beasts had to be paid on a daily basis, although somewhat less when marching back to base without a load (*konghui*) or waiting for the next batch of troops. If the batches were too large, the danger of insufficient transport could arise.

The logistics stations were supervised by a high-ranking member of the local administration, like district magistrates, sub-prefects or prefects. In some places, when the organizational work was too complex, the supervisor was assisted by civilian officials from a neighbouring district. Very often the organization of the logistics line was also laid into the hands of appointees (*houbu*) who did not yet occupy a post and were serving on probation. Other persons engaged for work in the logistics line were suspended officials observing the period of mourning, or officials that had been dismissed because of some offence. Especially the appointees and the degraded could benefit from the logistics duty because it offered them a chance to prove competence and eventually to be appointed to an office or reemployed, respectively. Yet the government also profited from the situation of these persons by using them as unpaid officials. Inside a province, the highest supervisors of the logistics route were the governor-general, the governor, the provincial administration commissioner and the provincial surveillance commissioner, i.e. the highest civilian authorities. In the case of Banner troops marching to the front, the provincial Banner general was also involved. The transport of rice and equipment was organized by a general transport bureau (*zhuanyun zongju*) that supervised all logistics stations. The procurement of food was the duty of the purchase-and-transport bureaus (*caiyunju*).[8]

Food for the troops could be organized in three different ways. Most efficient for the troops was the handing out of daily rations prepared by the logistics station. For this purpose the local government had to purchase and prepare rice and food, tea and firewood to be handed over to each individual or tent crew as soon as they reached the territory. The common troops then cooked their meals themselves or had them prepared by their manservants (three manservants for ten common Green Standard troops).[9] This method saved the troops time and prevented contact between the local population and the army. It was seldom used, though, because the troops complained that they were not given sufficient food, while the money earmarked for the purchase of food might be

[8] Lai, *Qianlong zhongyao zhanzheng*, p. 200.
[9] *Hubu junxu zeli*, juan 3, fol. 4a.

embezzled by the district magistrate. The most common method was to hand out money to the troops with which they purchased rice, vegetables and meat from sutlers. These sutlers had to be invited first to come to the camp, where they were allowed to sell food in special booths built for this purpose. This way a small market street (*maimaijie*) came into being at the camp site.[10] Contrary to the first method, in this case there was no fear of embezzlement on the part of the troops, but there were often disputes about the price of items the sutlers sold, or merchants felt badly treated by soldiers. In any case, these market streets made for some change during the march to the war theatre, and the troops had also the chance to purchase objects or services beyond the basic requirements of diet. They also saved the local government considerable organizational work because it was easier to hand over money to the troops instead of organizing the purchase of food and fuel. In rare cases, when there was no opportunity to provide food to the troops on the march, especially on forced marches or on marches outside provincial territory where no logistics stations had yet been established, troops were handed over several daily rations to carry along. There was a special regulation for Banner troops from the three northeastern provinces that were handed over the victuals for several weeks before they reached Beijing. Troops from Heilongjiang were given rations for 40 days, those from Jilin for 30 days, and troops from Shengjing (modern Liaoning) for 15 days. It can be seen that in practice this method was not restricted to the northeast Banner troops. In Europe, this method seems to have been rarely used because it was much more densely populated and therefore provided more opportunities organize food from local markets.[11] Even in China, the method to equip troops with rations for several weeks was only used for forced marches in sparsely inhabited territory.

Troops were normally given a fixed amount of money to buy rice (*kouliang yin*) and a certain amount to purchase other food with (*yancai yin*). They were often handed over the money in silver and had first to change this into copper cash before meeting the sutlers. Yet it was also possible that the local government took over the task to change the money in advance and handed over copper cash to the troops, which saved them a lot of time. It must be assumed that the silver was, although meant as

---

10 Dai, "Nourish a Strong Military", p. 89. Dai, "Military Finance", p. 303.

11 One such instance is mentioned in van Creveld, *Supplying War*, p. 29. The logistics problems in Russia are described on pp. 63 ff.

a pay for daily rations, not paid out day by day, but in a lump sum that was sufficient to cover the needs for a certain time. The soldiers liked to be paid out in cash rather than in silver because the official exchange rate was 1,000 cash against 1 *liang* of silver, while the market exchange rate was, at least until the end of the Qianlong reign,[12] about 800 cash against 1 *liang* of silver, which meant that the government gave a clearly better rate of exchange than the money-changers in the camp. In the beginning, when there was still no regulated accounting, there were many incidents that soldiers claimed their rations twice. The bill for that had in the end to be footed by the civilian official supervising the logistics station.[13]

Troops were normally paid out money (*zhese* 'converted commodity') inside the provincial territory, while beyond the border rice was given (*bense* 'original commodity'). There were also mixed forms in which half the daily provisions were paid out in money and the other half was given in rice. The normal daily ration was 1 *sheng* of rice (1035 ml), or, alternatively, 1 *jin* (597 g) of flour. The War Expenditure Code from 1785 reduced this quantity to 0.83 *sheng* (859 ml).[14] Troops were, furthermore, entitled to be given rice and could not be forced to accept money instead of rice. Instead of rice they could also be given a fixed quantity of flour and meat. For both flour and meat of different kinds there was a prescribed exchange rate between the various kinds of food, for instance, six sheep served as an equivalent to one cow or 30 *sheng* (31 l) of rice, as already explained in the first chapter. During the second Dzungar campaign the troops were given dropped horses, or mares and foals that could not be used as beasts of burden.[15]

The regulations for war expenditure include a paragraph fixing the cost for renting rooms from a private person, a monastery or a hostel. It seems that at least for officers is was not uncommon to spend the one or other night on the march to the battlefield in a private billet instead of in a tent. The price was fixed at 0.3 *liang* for a small room and 0.4 *liang* for a larger room.[16] It is not clear if monasteries were to be paid for providing accommodation. In case that cantonment was taken in private

---

12 Glahn, *Fountain of Fortune*, p. 210. Peng Xinwei, *Zhongguo huobi shi* (Shanghai: Shanghai renmin chubanshe, 2007; originally published in 1954), p. 605. Vogel, "Chinese Central Monetary Policy", pp. 5–7.

13 Lai, *Qianlong zhongyao zhanzheng*, p. 58.

14 *Hubu junxu zeli*, juan 2, fol. 1b.

15 Lai, *Qianlong zhongyao zhanzhzeng*, pp. 200, 217. See also *Jinchuan dang* 39/III/00301–303 (QL 39/9/14).

16 *Hubu junxu zeli*, juan 9, fol. 7b.

households, the inhabitants were often reaccommodated. The advantage of cantoning soldiers in private houses was that they lived closer to shops where they could procure their own food, yet the disadvantage was that it is not easy to control troops that are scattered over various quarters of a town. The simplest form of accommodation were thatched huts that were constructed in the open plain. Even for such a simple kind of shelter for the night, regulations were made concerning the cost.[17]

Camps were strictly guarded day and night, and it was theoretically impossible to penetrate into a camp without being challenged by the guards. Neither were civilians allowed to enter the camp, nor could soldiers leave it without permission. In the face of such a regulation it is quite unclear where the foodstalls were to be set up. It might therefore be that the strict regulations for the security patrol and the entrance guards were only adhered to when Banner troops from the northeast came to Beijing or when troops camped near larger cities. Reports from the war area show that it was very common for sutlers or monks to more or less freely enter the camps.[18]

Common troops and officers on the march were to be served by people hired from among the local population. During the whole campaign they were supported by manservants (*genyi*) that were hired in the area of the home garrison, yet there was also a large group of personnel hired in the logistics stations on the march to the front. As opposed to manservants these temporary servants (*congyi*) had only to do work for the troops during the time they passed through one district. Concerning the manservants, there were also clear regulations by how many persons an officer of a certain rank and type and troop was to be waited on, while no such regulations existed for the temporary servants. Most important for the transport of military equipment and food were the owners of carts, mules and horses—or, in extremely rough territory, porters—that were in the beginning paid different prices in each province. Also the pay varied according to the objects they transported (rice was paid according to different criteria from military equipment or cannons).[19] From 1778

[17] *Hubu junxu zeli*, juan 6, fol. 5b–6a. *Jinchuan junxu li'an*, juan 2, fol. 71b.

[18] *Jinchuan dang* 38/III/00207 (QL 38/7/22). During the first Jinchuan war natives were therefore easily able to conduct espionage in the camp.

[19] *Jinchuan junxu li'an*, juan 2, fol. 43b–44a. *Hubu junxu zeli*, juan 5, fol. 1a, 2b. In the regulations the pay was the same for 1 *dan* (a volume measure) of rice as for 130 *jin* (a weight measure) of equipment. A deviation from this thumb rule was that each porter was to carry 50 *jin* of equipment in mountainous territory, while it was expected that a man was to carry half a *dan* of rice (logically corresponding to 65 *jin*).

on their pay was standardized. When porters or cart owners served outside provincial territory they were given a higher pay than when serving inside their home province. Before the War Expenditure Code was issued the prices differed in all provinces. In order to cut cost it was decreed that the regular labourers of the courier stations had to be used first to transport the baggage of the troops, and only if their number was insufficient labourers could be hired from among the local population.

Similarly, the use of horses owned by the local garrisons as riding animals, draught animals and beasts of burden was preferred. Only when their number in the garrison or courier stations was insufficient private animals could be hired or purchased. In order to cut cost the government issued the decree that carts were the preferable means of transport for bulky goods. Only where the terrain prohibited the use of carts or where carts were not available mules could be used.[20] The most expensive means of transport, that by human porters, was accordingly the least preferable method and was only allowed in extremely difficult territory where mules were not able to walk or were not available. There were even prescriptions that one cart had to transport the luggage of six men, what load a mule had to carry, and what amount of rice had to be transported by a porter.[21] Prices for hiring boats to cross rivers or to transport troops along a waterway were likewise fixed. There were, for instance, higher prices to be paid to the towers when travelling upriver than when travelling downriver.[22] River transport was an important issue in both China and Europe because it was often cheaper and much less stressful for moving cumbersome or very heavy objects, like artillery.[23]

The logistics lines inside provincial territory were more or less identical to the courier routes. The logistics stations imitated the structure of the courier stations and made use of their staff, animals and equipment. During the war, couriers of course used the regular postal stations, but high officials also profited from the system of the courier stations to quickly travel from the war theatre to the capital. The cost for the logistics stations in provincial territory was thus covered by the regular expenses of the courier stations, while additional cost that accrued by hiring carts, horses or manpower, was allowed to be accounted for as war expenditure.

---

20 *Hubu junxu zeli*, juan 5, fol. 1b, 2b.
21 *Hubu junxu zeli*, juan 5, fol. 1b–3b.
22 *Hubu junxu zeli*, juan 5, fol. 2a–2b.
23 Van Creveld, *Supplying War*, p. 10. Barthorp, *Marlborough's Army*, p. 15, says that each cannon, except the light three-pounders, was drawn by a team of eight to ten horses.

The obligatory involvement of regular state institutions and civilian officials in the organization of warfare lowered cost considerably but also to a certain extent hampered a professional approach and the development of a kind of general staff that included officers especially familiar with the problem of supplies. The organization of logistics was put on the shoulders of civilian officials that were not given a higher pay for this additional duty carried out for the military. Much more exploitative was the hiring of candidates for office or dismissed officials that had to work for nothing. Last but not least the commandeering of manpower, waggons and animals from the locals were of course not very popular, even if the government was willing to pay for them. The Qing government was well aware of the burden inflicted on the population and therefore had the various troops after the war often marched back along a different route from the one they had taken to the war theatre.[24]

Outside provincial territory the supervisors of the logistics stations were either dispatched from administrative units in a nearby province or came from among the many candidates for office. The logistics stations between the external border of the provinces and the war theatre did not only serve to take the troops to the enemy, they also served as something like an umbilical cord between the war-waging province and the army on the move and its camps, respectively. The two Jinchuan campaigns were extraordinary cases with view to the mobility of the army. While troop movements in the Dzungar campaigns played an important role, in Jinchuan there was virtually no movement at all for many months. The logistics lines in Jinchuan therefore had a less temporary character than during other campaigns. All foodstuff and equipment transported along them was forwarded according to a relay system ('relay transport' *gunyun*). Labourers transported the baggage from one station to the next and then returned to base. Alternatively, there was also the method of long-distance transport (*changyun*) that saved the cost for logistics stations. The carriers in this case travelled for several days before they delivered their load and returned to base. Because it was not possible for loads to be immediately reloaded, each station disposed of granaries, sheds and storehouses where rice and equipment were kept temporarily, and of shacks where the porters slept. Transport vehicles, animals and transport

---

[24] During the Gurkha campaigns, for instance, the northeastern Banner troops returned via the Qinling Range between Sichuan and Shaanxi, while they had crossed the Qinghai plateau on their way to Tibet. See Lai, *Qianlong zhongyao zhanzheng*, Map 6 in the appendix.

tools like sacks or baskets were owned either by private persons or by the base station. Each 'grain station' (*liangzhan* or *liangtai*) therefore disposed of labourers weighing, checking and repacking the rice. It was prescribed that granaries and storehouses had to be equipped with mats to prevent moisture from getting into the rice when lying on the ground.[25] It was also prescribed that sheds for the labourers had to be erected and that bamboo and wood were to be used as the preferred building materials for granaries and other buildings of these temporary logistics stations. Because they were a target for the enemy, they were protected by 'stout men' or armed labourers.[26] While inside provincial territory the courier stations often served as logistics stations, the relation was somewhat inverted outside provincial territory, where logistics stations also served as courier stations. The labourers in the logistics stations were recruited from among the population or provided by native chieftains. The Mongols called these native labourers *ulaa*, a term that originally designated courier horses or the courier system itself. The term *ulaa* was also used for long-distance transport (*wula yun*) outside Mongolia, especially in regions outside provincial territory.

Recruited Chinese labourers ('labourers from the villages' *lifu*) had according to the regulations to be sent home after three months of service, later on after five months.[27] They were granted a pay and given daily rations of rice. Fare more comfortable for the government were so-called 'guest labourers' (*kefu*) or 'native labourers' (*manfu*) that were not recruited but flocked of their own accord to the stations to offer their services. The government did not only have to feed its soldiers and the civilian officials involved in the campaign, but also to care for the daily rations for up to one thousand labourers per station. In rare cases—during the two Jinchuan campaigns and the second Gurkha war—the government commissioned private merchants to transport rice to a certain destination. This kind of transport was called 'merchant transport' (*shangyun*) in contrast to the normal 'official transport' (*guanyun*). For the labourers employed by these merchants the government did not have to provide daily rations, while for recruited labourers as well as for 'guest labourers' daily rations were to be provided as well as an allowance called 'family

[25] *Jinchuan junxu li'an*, juan 2, fol. 60a.

[26] Lai, *Qianlong zhongyao zhanzheng*, p. 57. *Jinchuan junxu li'an*, juan 2, fol. 71a.

[27] *Hubu junxu zeli*, juan 6, fol. 3a–3b. *Pingding Liang Jinchuan fanglüe*, juan 92, fol. 8b–9b (QL 39/3/xinwei). *Jinchuan junxu li'an*, juan 2, fol. 66a, 70b–72a. See also Dai, "Qing State, Merchants", p. 49.

comfort' (*anjia yin*).[28] Merchant associations that donated contributions to the war expenditure were also allowed to directly deliver rice to the camps. They cared for the transport by themselves.

The rice relayed on along the logistics stations came from the regular 'ever-normal' district granaries (*changpingcang*), from the donation granaries (*changjiancang*), where rice was stored that had been donated as a contribution to academic institutions, or from the semi-private 'public spirit' or 'charity' granaries (*yicang*) that normally provided disaster relief, or it was purchased on the market.[29] Finally, there was also rice donated during contribution campaigns.[30] Rice as booty or food presented by native allies played only a marginal role because it could only accidentally be acquired and was not subject to planning. In Eastern Turkestan the grain harvested in the military agro-colonies (*tuntian*) also served to feed the army in case of military campaigns.

The production of military equipment was subject to regulations of the Ministry of Works. During the war, when weapons had to be produced, repaired or replaced or consumables as gunpowder, bullets or arrows to be provided, the home garrisons or the garrisons of the war-waging province made these objects and had them sent to the war theatre. Weapons were also produced by the Ministry directly, especially such that required high technical skills like bows and quivers. Simple weapons like sabres could be purchased on the market. The same goes for nitre and sulphur for the production of gunpowder, if the stocks in the provincial garrisons were emptied. Cannons could be either transported from the garrisons to the battlefield, or were cast in the immediate vicinity of the camps, as happened during the Jinchuan and the Myanmar campaigns.[31]

It was common that each garrison disposed of certain amounts of weapons and consumables. When marching to the battlefield each soldier carried with him 1 *jin* (597 g) of gunpowder and 40 bullets.[32] In the war zone a soldier had on average to carry a load of thirty to forty

---

[28] Lai, *Qianlong zhongyao zhanzheng*, p. 203, says that no family allowance was to be paid for guest labourers, yet an edict quoted in the *Pingding Liang Jinchuan fanglüe*, juan 106, fol. 10b (QL 39/9/xinwei), proves the contrary.

[29] Concerning the *yicang* granaries, see Will, *Bureaucratie et famine*, pp. 176–180. The translation 'charity granaries' follows Will and Wong, *Nourish the People*.

[30] During the second Jinchuan war, for instance, merchants contributed 1.1 million *liang* for the purchase and transport of rice to the camps. *Gaozong shilu*, juan 944, fol. 18b–19b (QL 38/10/gengyin). Also documented in *Jinchuan dang* 38/IV/00041 (QL 38/10/6). *Gaozong shilu*, juan 956, fol. 8b–9a (QL 39/4/wuzi).

[31] Lai, *Qianlong zhongyao zhanzheng*, p. 295. *Gongbu junxu zeli*, juan 1, fol. 4a.

[32] Lai, *Qianlong zhongyao zhanzheng*, p. 293.

jin (about 18 to 24 kilos), of which gunpowder, 200 bullets, fuses, his musket and dagger alone counted for more than 30 *jin*, plus rations for ten days, which weighed in at more than 10 *jin*. When adding tent and cooking woks to that the weight of the burden for each soldier was probably more than 30 kilos.[33] Boots, winter clothes, tents and cooking pots were also produced in the garrisons and transported to the camps. Sometimes other government agencies like the Imperial Household Department (*neiwufu*) provided equipment like cotton corslets during the first Jinchuan war, and the provinces of Hubei and Hunan produced winter cloaks for their troops during the second Jinchuan war.[34] The brass needed to cast cannons was borrowed from the provincial mint that was in possession of large amounts of metals anyway. In rare cases protective clothing or horse covers were donated by private persons within the framework of contribution campaigns.[35] Of course weapons and equipment could also be acquired as booty from the enemy.

There was never a standardization of weaponry in Qing China so that each province issued its own rules for the features and production of weapons and military equipment. Even the central government itself possessed its own standard code that only applied to equipment owned by the Imperial Household or the Banner units in Beijing. As it therefore was not possible to equip the whole army with cannonballs or bullets of a standard size, these items had to be produced on site and according to need. Even the rules for gunpowder differed from province to province, as could be seen in the chapter on expenditure. Another problem was that a considerable part of the clothing and weapons was, according to the regulations, to be organized by the troops themselves.[36] This means that at the beginning of a campaign the soldiers first of all had to purchase suitable boots and uniforms as well as short-range weapons like daggers and swords. These objects were purchased on the local market, a custom resulting in very different types and formats. The government did also not care for the repair and upkeep of these weapons and uniforms, so that a

---

[33] *Pingding Liang Jinchuan fanglüe*, juan 85, fol. 7b–8a (39/1/jiwei); juan 86, fol. 20a–20b (QL 39/1/renshen).

[34] Lai, *Qianlong zhongyao zhanzheng*, pp. 293–294.

[35] *Jinchuan dang* 38/IV/00315 (QL 38/12/16). See also Lai, *Qianlong zhongyao zhanzheng*, pp. 418–421.

[36] Lai, *Qianlong zhongyao zhanzheng*, p. 296. Dai, "Military Finance", p. 302. The Military equipment regulations from 1791, *Junqi zeli*, fol. 16, for instance, say that bows, arrows, quivers and all clothes and boots were to be organized and paid by the troops themselves, e.g. Green Standard garrisons in the province of Zhili.

lot of units of the Qing army were often poorly equipped with useless or antediluvian pieces. Firearms appear to have been provided by the garrisons, yet the simple construction of matchlock muskets allowed them to be produced by private persons in any case, even if it was officially forbidden for commoners to own and to use firearms. In many instances protective clothing was presented to troops as a kind of reward for heroic fighting, which shows that cotton corslets or even winter cloaks did not normally belong to the equipment of troops.

The transport of military equipment to the camps was effected along the same pattern as grain, and the pay and daily rations given to porters were largely the same as for the transport of grain. There were special regulations for the transport of cannons in mountainous regions.[37]

Very important for the whole progress of the war was the participation of craftsmen to take care of the construction of bridges, palisades, boats, the production of nails and other iron parts, or the fabrication and reparation of weapons. Each garrison disposed of a certain amount of craftsmen that accompanied the army to the field, but there were a lot of workmen to be recruited from the population. The supervisors of the logistics also had to see to the recruitment of suitable personnel. Especially noteworthy are the cannon founders employed during the Jinchuan and Myanmar campaigns. It is not known where these experts came from, but it must be assumed that they were either employees of the provincial mint in Chengdu or Kunming, Yunnan, respectively, or were working in the Firearms Brigade (*huoqiying*) in Beijing.

In the following paragraphs a few important examples will show how logistics was implemented, and what problems were met in particular cases. The respective information is based on the detailed studies of Lai Fushun on war expenditure and logistics (*junxu*) during the Qianlong reign. These three descriptions will be enriched by a detailed study of logistics matters of the second Jinchuan campaign which provided a testing zone for perfection and can be seen as the conclusive factor for the compilation of the War Supply and Expenditure Code *Junxu zeli*.

Mismanagement During the First Jinchuan Campaign

The first Jinchuan campaign (1748–1749) was fought in the mountains of Eastern Tibet, in the western outskirts of the province of Sichuan. There

37 *Hubu junxu zeli*, juan 5, fol. 2b–3b, 3b–4a. *Jinchuan junxu li'an*, juan 2, fol. 66a–66b, 67a–67b.

were 70,000 troops participating in the war, 13,000 of which came from Green Standard garrisons in the provinces of Shaanxi, Gansu, Yunnan and Guizhou, and a further contingent of 15,000 troops had been sent from Shaanxi and Gansu. 5,000 Banner troops marched from Heilongjiang to Sichuan, 2,000 Banner troops came from garrisons in other provinces, and 1,000 from the Banner garrison in Chengdu.[38] There were also troops from Hubei and Hunan. The longest distance was covered by the troops from the northeast. They had to march to Beijing and then to cross the provinces of Zhili (modern Hebei), Henan and Shaanxi before entering Sichuan. There were eight large logistics stations in Zhili and Henan, and seven in Shaanxi. This means that the troops could only change their horses every second day. For each station about 800 transport animals and 300 carts were necessary to transport the baggage and the troops, yet in the end, Shaanxi alone needed 13,000 transport animals because the route between Shaanxi and Sichuan crossed the Qinling Range where large parts of the route that led along river gorges consisted of plank road viaducts (*zhandao*) impracticable for carts. In Sichuan, the final destination, two problems were faced in the logistics stations inside the province, namely lacking manpower, because many labourers had to be used as porters in the mountains of Jinchuan, and an insufficient amount of transport animals. It was therefore necessary to bring in beasts from Shaanxi, Hubei, and also from Shanxi, which province was not touched by the troops. Much more interesting is that 3,000 animals were procured from Beijing, yet not directly, but in a relay system in which the mules from Beijing were taken to the next station, the mules from that station again to the next, and so on.[39] The amount of horses needed per person ranged from three animals per man to three per two men. The latter method was finally preferred, and it was ordered to save baggage by dispensing with surplus tents and cooking utensils. For such a high amount of horses considerable quantities of fodder had to be procured. This was mostly done by supplying the stations with various kinds of beans from the district granaries. The reason for such a huge amount of animals needed for transport was that the emperor had ordered to march in large batches of 400 or 500 troops, in order to arrive faster in the war zone. The frequency with which these batches left their garrisons was too high. Instead of less strenuous intervals of five days, many contingents started their march

[38] Lai, *Qianlong zhongyao zhanzheng*, p. 65.
[39] Lai, *Qianlong zhongyao zhanzheng*, p. 66.

with an interval of only two or three days, so that the animals could not rest or they even returned too late to their base for the next batch of troops. The lack of transport animals even resulted in the offer to hand over 15 *liang* to each soldier with which to purchase a horse of their own, which in the end proved impossible because there were no transport animals left around the logistics stations.[40] This procedural method was not at all extraordinary but found its way into the War Expenditure Code.[41] The commentary to this paragraph declares that the precedents from the Jinchuan campaigns were the sole instances that troops were only given a riding horse and had to dispense with additional transport horses or horses for the manservants. In the end the troops could not but march on foot. Even the highest commander, Fuheng, reportedly dispensed with his mount and, as a respectable officer, shared the fate of his troops.[42] For the march back there was a suggestion to have the troops from the north-east shipped down the Yangtse River and along the Grand Canal back to Beijing, but in the end it was decided to have them march back in order to save time.

The many troops participating in the first Jinchuan campaign were supported by almost a quarter of a million labourers. All these people had to be fed with rice, yet the state granaries in the province of Sichuan had only grain reserves of 400,000 *dan* (41,400 m^3), so that it was necessary to purchase grain on the market and to prohibit the export of rice to the provinces of Hubei and Hunan, at least for the duration of the war. The price for grain on the market was between 0.6 and 0.9 *liang* per *dan*, yet the transport conditions along the steep mountains paths inflated it to 12 or even 16 *liang*.[43] A smaller amount of rice originated in contributions by private persons or merchant associations, yet there was considerable embezzlement of funds and grain originating from this source, leading to the decision to halt the contribution campaign for a while. The troops were additionally fed with roasted flour and sometimes given sheep or oxen. As is already known, the rules allowed that these items could be substituted and the troops be given money instead. The dense network of logistics paths through the mountain valleys was originally not yet as

40 Lai, *Qianlong zhongyao zhanzheng*, p. 68.

41 *Hubu junxu zeli*, juan 8, fol. 1b–2b. The price quoted here is 6 *liang* and not, as told by Lai, 15 *liang*, yet a price of 15 *liang* can be confirmed in *Pingding Jinchuan fanglüe*, juan 17, fol. 6b (QL 13/11/guiyou).

42 *Pingding Jinchuan fanglüe*, juan 20, fol. 3b (QL 13/12/renyin). The incidence is even mentioned in Fuheng's biography in the *Qingshigao*, juan 301, p. 10446.

43 Lai, *Qianlong zhongyao zhanzheng*, p. 205.

sophisticated as later on. Thus in several instances the camps ran out of food or the rice was delivered to the wrong destination.[44] Granaries sometimes burnt down with the flames devouring the urgently needed food.[45] The lack of rice in the camps was partially also due to widespread desertion among the peasants that were recruited to do the hard work of transporting the rice on their backs. That work could be quite demanding for people not accustomed to it and resulted in quite a few dead among the recruited porters. Lai Fushun says that therefore half of the native population (or at least the males) were recruited to do porterage service because they were better trained in transporting heavy loads on mountain trails.[46] The government had the duty to pay out indemnities to the families of labourers that died during their work for the government. Fortunately a lot of rice was left over after the abrupt end of the campaign, and it was decided to donate it to the native chieftains or the families of perished labourers. The complexity of recruiting peasants to carry rice and equipment and the high responsibilities to be borne by the government could be somewhat resolved when it was decided to have part of the rice transported by commissioned merchants. These hauliers, either official merchants (*guanshang*)[47] or private merchants, were ordered to transport a certain amount of rice to a defined destination at a negotiated price, with the obligation to assemble porters and to purchase rice on the market.

With view to all the difficulties faced during that campaign it must be admitted that the supervisors of the logistics routes, the governors-general of Zhili, the governor of Henan and the provincial administration commissioner of Shaanxi, achieved a relatively good result when considering that only six to ten days were left between the imperial decree to set the troops moving and the day when their march began. Most disastrous was the logistic organization in the province of Sichuan where numerous high civilian officials were dismissed or even punished with the cangue for their failure to organize a sufficient number of horses, carts and labourers and

---

[44] *Pingding Liang Jinchuan fanglüe*, juan 59, fol. 13a–14a (QL 38/4/xinhai).

[45] Lai, *Qianlong zhongyao zhanzheng*, p. 210, based on *Gaozong shilu*, juan 337, fol. 7b (QL 14/3/bingyin). There were also some incidents when grain went lost in an incendiary during the second Jinchuan campaign, see *Jinchuan dang* 38/III/00448 (QL 38/9/1), 00453 (QL 38/9/1); 38/IV/00289 (QL 38/12/12), 00329 (QL 38/12/20), 00341 (QL 38/12/24).

[46] Lai, *Qianlong zhongyao zhanzheng*, p. 210.

[47] See Lai, *Qianlong zhongyao zhanzheng*, p. 207. 'Official merchants' were merchants that had a licence to purchase, distribute and sell scarce goods the production of which was mainly in the hands of government agencies.

the resulting chaos in the grain supply.[48] The overhasty planning for this campaign and the unrealistic expectations of the young, yet ambitious Qianlong emperor served as a negative example for future wars.

Long-Distance Logistics in the Western Campaigns

The Kangxi emperor led three campaigns against the Dzungar leader Galdan (r. 1671–1697). During these years the military commanders as well as the highest provincial civilian officials learned a lot about supplying an army that was campaigning far beyond the borders of 'civilized land'. The first campaign was undertaken in 1690 and involved 60,000 troops that marched along three routes to the west.[49] They took with them the rations for four months, although the Emperor expected a blitzkrieg campaign not to last longer than two months. The reason for this over-optimistic assessment was Galdan's own situation whose troops were nearly starving at that moment and began to eat their horses in order to survive. He nevertheless possessed a lot of fowling pieces (*niaoqiang* muskets) as well as cannons. The imperial army, too, took with them cannons, at least in the division personally commanded by the emperor. Yet already one month after the onset of the campaign the horses of the Manchus and their 10,000 allies from the Khorchin (Qorčin) Mongols were tired and rations scarce. The divisions could only advance very slowly because they had to wait for the supply trains, which gave Galdan the chance to withdraw. The Khorchin troops refused to pursue him because of their critical condition of supplies. In late August, two months after the beginning of the campaign, the Emperor likewise withdrew his contingent, after the famous battle of Ulan Butong where the Qing cannonaded Galdan's camel corral.[50] Only 400 troops remained in the west to ensure further negotiations with Galdan. The lesson from this first attempt at subduing the

---

[48] Lai, *Qianlong zhongyao zhanzheng*, p. 71. Dai, "Qing State, Merchants", p. 48. The cangue was a square wooden block worn around the neck in which the delinquents were presented publicly, much similar to the pillory known in the West. In this case, a circuit intendant was dismissed and punished.

[49] Perdue, *China Marches West*, p. 152.

[50] The armies of Nādir Schāh (r. 1736–1747), ruler of Persia, used camel-guns (*zanburak* 'little wasp'). They were mounted on camelback on a wooden frame with a swivel top. For them to be fired the camels had to kneel down. The battle of Golnabad in 1722, with the "intelligent" and "innovative" use of camel-guns as a weapon supporting the traditional lance and sword fighters, marked the end of an era in the style of warfare for Persia and Afghanistan. Compare Michael Axworthy, *The Sword of Persia: Nader Shah, From Tribal Warrior to Conquering Tyrant* (London, New York: Tauris, 2006), pp. 48–49, 74 (illustration).

Dzungars was that the "brute facts of distance, food, and horses made [...] the elimination [of Galdan] impossible".[51]

These problems in supplying a large army with grain and fodder for the horses were only overcome in the mid-eighteenth century, when the economic development made it possible to stock larger amounts of grain and to spend more money, and when the foundations of a logistics network were already laid. The territory to be crossed lacked water and grass, and there were no animals to hunt for additional food. From time to time heavy rainfall made the progress of carts, and especially the transport of heavy cannon, impossible. Matters were aggravated by serious embezzlement cases with the supplies that were, after all, to be attributed to the imperfect organization. High-standing officials illegally advanced two months of extra rations to the army of Shengjing (modern Shenyang, Liaoning) and five months of additional salary just because the officially allowed rations did not meet the needs of the troops.[52] General Feiyanggu sold his stocks of tea and cloth to buy horses and grain for his troops.

Some troops sold their weapons to the Mongol tribes in the vicinity in exchange for food, but on the other hand, the Mongol allies were in the beginning prohibited from purchasing grain and horses on the local markets, for fear of shortness and skyrocketing prices. Before the campaign and thereafter, riots among the Bannermen threatened the peace of the Capital, firstly because the prices for horses had almost been doubled by imperial command, in order to protect local people from exploitation, and secondly because a large number of troops was not able to repay their debts that had accumulated as baggage pay to prepare their equipment. The emperor decided to take over the debt which amounted to 16 million *liang*. This has to be added to the 3 million *liang* that the campaign had officially cost, a sum that corresponds to less than a tenth of the actual treasury holdings of 50 million *liang*.[53]

The second campaign, undertaken in 1696, likewise involved three divisions, with a total number of 70,000 troops. They took along the supplies for 8 days, while supplies for a further period of 50 days were to be shipped soon.[54] The provinces of Zhili, Shandong and Henan were to provide 1,300 carts, and the governor-general of Zhili had 6,000 carts built for the army, each of them able to carry 6 *dan* (621 l) of grain. The troops from

[51] Perdue, *China Marches West*, p. 159.
[52] Perdue, *China Marches West*, p. 159.
[53] Perdue, *China Marches West*, pp. 160–161.
[54] Perdue, *China Marches West*, pp. 181–182.

Shaanxi were equipped with 22,400 *dan* (2,318 m^3) of grain which was thought to suffice for five months. Part of this amount was substituted by cattle, and in order to facilitate transport, each soldier carried 5 *sheng* (5.17 l), as well as a certain amount of silver to buy food with in the stations in the province of Gansu. Horses were purchased from the Khalkha Mongols: 1,000 in Guihua (Hohhot, Inner Mongolia), and 2,000 in the Ordos region from the Mongol Khorchin banners. The province of Fengtian (Liaoning) supplied 300,000 bundles of hay to feed the horses.[55] The imperial army was well equipped with gunpowder and disposed of 235 large cannon as well as of 104 smaller pieces. The latter could be transported on camelback and were taken to the war theatre, critically slowing down the advancing army. The heavier pieces, weighing between 8,000 and 10,000 *jin* (4.7 to 6 metric tons), had to be left behind. Heavy rainfall in the early months of spring made the roads impassable, but in some places it was possible to construct practicable roads with the help of willow twigs.

The spring campaign came to a halt, and the whole army waited for fresh supplies. Yu Chenglong, governor-general of Zhili, returned to the capital to collect 27,000 *dan* (2,795 m^3) of grain. A further 70,000 *dan* (7,245 m^3) were borrowed from the granary in Guihua, the equivalent of 20 days' rations for the troops.[56] The Ordos region became the most important source for horse fodder and venison as an additional source of food.

The last campaign against Galdan was begun in February 1697. The supply officers had in the meantime accumulated sufficient experience in organizing carts, horses, fodder and grain. The surrender of the oasis of Hami and of the region of Kokonor (Qinghai) made it much easier to procure supplies. Hami was the easternmost Uyghur city in the western region, and can be seen as a first stepping stone for the territorial conquest of Eastern Turkestan. Ningxia on the western edge of the great Yellow River Bend was for the first time used as a source for supplies. 3,000 *dan* (310 m^3) of grain were shipped upstream and were to supply the vanguard with 45 days of provisions, a timespan that would get them to the Altai Range, the heartland of Galdan's empire. It was not only Yu Chenglong who supervised the whole logistics system this time, but the Kangxi Emperor himself became a "supply sergeant", as Perdue says.[57]

55 Perdue, *China Marches West*, p. 183.
56 Perdue, *China Marches West*, p. 195.
57 Perdue, *China Marches West*, p. 201.

Imperial logistics this time had been so much superior over Galdan's narrow supplies that it was said Yu Chenglong had destroyed Galdan, and not the supreme fighting power of the Manchu troops.[58] Persons later remonstrating against the high expenditure for logistics were criticized by the Yongzheng Emperor with the words that he was not only willing to spend 250,000 *liang* for supplies, but even several million, if necessary.[59] This politics of war expenditure foreshadowed the immense amounts of funds his son, the Qianlong emperor, would eventually disburse to feed his army in time and sufficiently.

In 1716, Barköl (Balikun) was established as a supply base even more to the west. The military agro-colonies (*tuntian*) in the far west were planned to bring relief to the problem of supplying troops manoeuvering a thousand miles from home, but in the end the 33,000 garrison troops in Barköl and Turfan consumed 6,690 *dan* (692 m^3) of grain monthly and could only subsist with imports from elsewhere. In this situation the colonies would barely have been capable of feeding an army marching as far as Ürümchi, where the Dzungars had withdrawn to. In spite of such shortcomings the decisive role that the issue of logistics played for the campaigns in the west and in Tibet could no longer be ignored. The organization of supplies also underwent certain political changes, as Peter Perdue says. While formerly the Manchus and Mongols used to fight, and the logistics was managed by Chinese officials, the campaigns of the Yongzheng reign saw a participation of Manchu officials also in the organization of supplies. The growing administrative machinery for the deliveries of grain and the organization of riding and transport animals inevitably also resulted in a larger amount of cases of embezzlement and extortion. Officials often short-changed common troops, and superior officials extorted money from their subordinates.[60]

The Qianlong emperor's first campaign against the Dzungars (1748–1750) was fought by 50,000 troops whose origin was very different from those of the Jinchuan war. Only 10,000 were Green Standard troops, about 7,000 were Banner troops, and the rest consisted of troops from various allied peoples like the Oirats, Khalkhas, Tümed, Daghur and Solun, or Uyghur troops from Hami.[61] The 20,000 troops garrisoned in the northeast and inside

[58] Wei Yuan, *Shengwuji* (Sibu beiyao edition), juan 3, fol. 37b, quoted in Perdue, *China Marches West*, p. 201.
[59] Perdue, *China Marches West*, p. 231.
[60] Perdue, *China Marches West*, p. 229.
[61] Lai, *Qianlong zhongyao zhanzheng*, pp. 76–77.

Chinese provinces marched along the so-called western route through the provinces of Shaanxi and Gansu and needed more than 60,000 animals that came partially from the home garrisons and were partially bought. Because officers used horses from the home garrisons it was sufficient to equip each logistics station in the province of Zhili with 400 horses. It was also decided not to care for rooms in hostels, but all troops took along tents to spend the nights. Even higher officers passed the night in tents. Hay and beans as fodder for horses were purchased on the market. This time there were not only large stations in the provinces of Zhili and Henan but also interstitial stations where horses were not changed. Apart from horses for riding, there were also about 250 carts assembled in each station to transport the baggage of five hundred troops.[62] In some places around the Hangu Pass in the province of Henan it had been necessary to widen the narrow paths through the mountains. The troops were paid out copper cash to buy food and fuel with. This cash was directly sent from the provincial mint (*baoshaanju*) to the logistics stations. In Shaanxi two different routes were chosen, the traditional southern route that had already been used during the Kangxi and Yongzheng reigns, and a new northern route running through dry and hilly lands. The southern route was much longer and included ten logistics stations, and there were few problems with commandeering horses, even for batches numbering 500 men that needed 1,250 horses. In the whole province of Shaanxi 3,000 carts and 12,500 transport animals were hired. Unlike in Zhili and Hebei, the troops lived in hostels when passing the province of Shaanxi, and so saved the cost of 4,900 *liang* that had accrued for the production and repairing of tents during the first Jinchuan war.[63] In Gansu the northern route was 28 stations long, the southern one 25. The province had to care for 42,000 transport animals. Most of them were bought on the market, which did not present a serious problem because the supply of horses in Gansu was much better than in the central provinces. The number of horses available was, nonetheless, insufficient so that it was decided not to change horses in each station, which made it possible to save an amount of 20,000 animals. Yet the long-distance transport also exhausted the horses, making it impossible to use them as war horses during the campaign. Six thousand horses were provided by the garrisons of Gansu,

62 Lai, *Qianlong zhongyao zhanzheng*, p. 78.
63 Lai, *Qianlong zhongyao zhanzheng*, p. 79.

while the balance of 20,000 had to be hired from the population.[64] In the eastern part of Gansu the Yellow River had to be crossed a second time. For this purpose the provincial administration commissioner in good time prepared a sufficient amount of boats to ferry the troops. Sutlers supplying the troops with food had to be commissioned early on to assemble in the stations and replenish the booths, because there were only a few larger towns in Gansu to purchase food from. Interestingly enough the troops faced trouble when a number of manservants deserted. A special force of military police had to be detached to arrest the deserters. Outside provincial territory there were much more serious obstacles, especially the lack of food and water. This was the region where during the earlier campaigns under the Kangxi and Yongzheng emperors temporary logistics stations had to be erected. At least, experience had been gained during these days when wells had to be sunk in advance and foragers to be hired who cut grass for the horses. It might be remembered that outside provincial territory it was not allowed to purchase fodder for the animals. In an effort to improve and simplify this extremely complex procedure, during the second Dzungar campaign that took place a short time later (1755–1757) only 2,500 troops were marched along the western route.[65] In Shaanxi and Gansu a critical examination of the topographical conditions by the provincial surveillance commissioner reached the result that the northern route might be chosen because it saved two days. In order to meet the demand for food and water, oxen and mules were bought to draw water from existing wells that was then transported to the logistics stations. In Gansu, more than 60,000 transport animals were prepared this time, a sixth of which came from the provincial garrisons. The need for such large numbers brought the province of Gansu to the limits of its supplies. In order to fill it the logistics managers had to pay stiff prices for horses and mules hired from private persons. Even the time spent in the stations while waiting for troops to arrive was paid for. This time, 1,000 horses had to be brought in from the province of Sichuan, a region normally not blessed with horses, as well as 5,000 from Shaanxi and several thousand camels from Shanxi.[66]

The 78,000 troops and civilian staff marching along the western route brought with them the provisions for four months, batches that started

---

64 Lai, *Qianlong zhongyao zhanzheng*, p. 81.
65 Lai, *Qianlong zhongyao zhanzheng*, p. 83.
66 Lai, *Qianlong zhongyao zhanzheng*, p. 85–86.

later were supplied with provisions for three months. Thus the earlier contingents could survive some time without supplies from outside. These provisions consisted of 11,200 *dan* (1160 m^3) of rice, 2.8 million *jin* (1,600 metric tons) of roasted flour, 750,000 *jin* (448 t) of plain flour and 20,000 sheep for the first contingents leaving the home garrisons. For those that marched later a further 18,000 *dan* (1860 m^3) had to be organized, as well as 4.2 million *jin* (2,500 t) of roasted flour and 1.3 million *jin* (776 t) of plain flour. In total, 100,000 *dan* (10,300 m^3) of grain were consumed by the troops on the western route, 15,000 *dan* (1,552 m^3) of flour, 120,000 sheep, and 22,000 *dan* (2,280 m^3) of beans as fodder.[67] Lai Fushun has calculated that the provisions taken along by the marching contingents of the western route weighed up to 120 *jin* (71 kilos) per man.[68] In the war zone the troops were fed with grain reserves from Hami, amounting to 34,500 *dan* (3,570 m^3), grain reserves from the garrisons in Gansu and Shaanxi, and rice reserves from the granaries in Gansu, whose huge store of 1.75 million *dan* (181,000 m^3) was more than enough. A very important type of additional food were dried fruit, dried meat, tea leaves, and a great number of live cattle. The town of Hami, for instance, provided 300,000 *jin* (179 t) of dried meat as well as 120,000 sheep. Tea leaves were transported in envelopes, and were so important for the nutrition of the army that 45,000 packages were delivered to the troops on the western route. During the second Dzungar campaign Hami supplied 81,000 *dan* (8,380 m^3) of rice, the town of Barköl 26,000 *dan* (2,700 m^3), and 120,000 *dan* (12,400 m^3) of rice were planned to be brought in from Shaanxi, yet this amount by far surpassed the need of the troops and their civilian supporters.[69] The garrison in Hami had been founded a short time ago so that in this hot spot region troops were ready when needed. The supplies consumed by this garrison amounted to no less than 40,000 *dan* (4,100 m^3) of grain per year. Though part of this was produced locally, still eighty per cent of their supplies were paid out in cash.[70]

The region of Qinghai delivered 11,000 foals and mares as food for the troops, and it was considered to import 4,000 oxen from Sichuan because Gansu did not have enough draught oxen. Yet as the animals from Sichuan

67 Lai, *Qianlong zhongyao zhanzheng*, pp. 216–217.

68 Lai, *Qianlong zhongyao zhanzheng*, p. 220.

69 Lai, *Qianlong zhongyao zhanzheng*, pp. 218–219.

70 Peter C. Perdue, "The Qing State and the Gansu Grain Market, 1739–1864", in *Chinese History in Economic Perspective*, ed. by Thomas G. Rawski and Lillian M. Li (Berkeley: University of California Press, 1992), p. 111.

would not be accustomed to the dry climate in Gansu, this idea was abandoned.[71]

The troops on the northern route were supplied with 35,000 *dan* (3,600 m³) of rice from the garrisons, as well as 6,000 *dan* (621 m³) of roasted flour. Part of the provisions had to be bought on the market at a convenient official price of 0.85 *liang* per *dan*. The transport in the desert, nonetheless, was too expensive so that in some instances each *dan* of rice had all in all cost 40 *liang* (normally up to 9.8 *liang*), while a price of 7.7 *liang* per *dan* had been the rule on the western route. The province of Gansu also served as a hub for the transport of no less than 220,000 packs of tea to the troops on the northern route.[72] The demands on the local grain market drove prices in the northwestern parts of the empire to unexpected heights, notably because the campaign coincided with a period of drought in Gansu. The logistics bureaus therefore bought grain on the private market or from neighbouring provinces, for which purpose the government had allotted 3 million *liang*.[73]

On the northern route from Beijing westwards, along which 30,000 troops had been dispatched to the war area during the first Dzungar campaign, camels served as the most important transport animal. They were not used in relay service but for long-range transport as far as Uliyasutai in western Mongolia. An amount of 11,000 *dan* (1,140 m³) of grain was transported by commissioned merchants at the rate of 2,000 *dan* (207 m³) that were shipped in intervals of 3 days. This relay transport method was very similar to that of the troops on their march to the war theatre.[74] This instance of commissioned private transport as well as others,[75] contradict the often-repeated argument that it was only the two Jinchuan campaigns and the second Gurkha war in the course of which the government relied on private entrepreneurs to have grain transported to logistics stations or camps.[76] Based on these precedents the War Expenditure Code *Junxu zeli*

---

[71] Lai, *Qianlong zhongyao zhanzheng*, p. 217.
[72] Lai, *Qianlong zhongyao zhanzheng*, p. 217.
[73] Perdue, "Qing State and Gansu", pp. 111–112.
[74] Lai, *Qianlong zhongyao zhanzheng*, p. 221.
[75] *Pingding Zhungar fanglüe*, *Qianbian*, juan 39, fol. 12b, 21a (YZ13/12/jiashan, yichou); juan 42, fol. 7a (QL 1/6/dinghai); juan 44, fol. 14a (QL 4/4/yiyou); *Zhengbian*, juan 2, fol. 15b (QL 19/4/bingshen). Merchants or owners of camel herds were apparently able to ship grain at a lower price than the government with its 'official camels' (*guantuo*).
[76] Lai, *Qianlong zhongyao zhanzheng*, p. 203.

exceptionally allows, in urgent cases to make use of privately organized transport.[77]

The campaign against the Muslim insurgents Burhān ad-Dīn and Khwāja-i Jahān (the two Khojas) in 1758–1759 was launched by 1,000 troops of the Vanguard Brigade (*qiangfeng ying*) of the Banners in Beijing, 2,000 Solun troops, 1,000 troops from Chahar, and 2,000 provincial Banner troops from Xi'an.[78] The troops from Beijing needed no more than one month to arrive in the war zone in the far west. The number of horses in the province of Gansu was this time so much reduced that 4,000 camels and 6,000 mules served to transport the baggage of the soldiers. Also a large number of carts and transport animals were hired from the province of Henan to facilitate transport. This time, the troops could also be fed by the grain produced in the military agro-colonies in Ürümchi, Ili, Pichan (Turfan) and Uchturpan (Uš) that provided more than 30,000 *dan* (3,100 m^3) of grain. Uyghur and Mongol allies also delivered contributions to the Qing troops, like the Prince of Alašan who presented 4,000 oxen and 14,000 sheep that could feed 20,000 men for as long as three months.[79]

In contrast to the poorly organized first Jinchuan campaign the experience gained in the earlier campaigns against the Mongols had excellent effects on the organization of the western campaigns under the Qianlong emperor. The preparations were begun a year in advance, thus there was adequate time to organize the necessary amount of food and transport animals. It was considered that in the densely populated provinces of Zhili and Henan the troops could stay in hostels or private houses, while in Shaanxi and Gansu it was better to prepare sites to set up about 280 tents per stations. The troops were paid cash to buy food directly from the many sutlers flocking to the camps. It was also decided that the troops were only given their daily rations and paid out the salted-vegetable pay, while during earlier campaigns it had been common in the one or other province to give the troops additional food that was paid for by the government. The method to have the troops take along the provisions for several months saved a great deal of organizational trouble. On top of this it was in the end probably cheaper to hire carts and transport animals for a lengthy period of time than to transport the troops and their baggage first and later the supplies. On the western route, in one incident during

[77] *Hubu junxu zeli*, juan 5, fol. 2a.
[78] Lai, *Qianlong zhongyao zhanzheng*, p. 86.
[79] Lai, *Qianlong zhongyao zhanzheng*, p. 218.

the first Dzungar campaign, the quantity of grain to be transported surmounted the transport capacities of the logistics stations by far. In another case the transport sacks burst and a lot of grain was lost. Among the animals taken to the camps it was not only the horses that died by exhaustion, but the mortality was especially high among sheep when driven too fast. In one instance, 15,600 out of 25,000 animals died.[80] Similar figures are also reported from the war in the West.[81] A large part of the camels brought in from the provinces of Zhili and Shanxi for transport likewise died of exhaustion.[82] It was therefore better to have them slaughtered and transport the dried meat to the camps, instead of taking live animals there. In no other war of that era, tea played a more important role than in the campaigns in the west. It might in many instances have been the only way to consume the water from the local wells, and it also supplied the troops with a certain amount of vitamins. The grain shipped to the west was supplied by many different sources, from the home garrisons or the provincial granaries to the time-tested institution of the military agro-colonies. This diversity spread the burden to deliver supplies to several parties of different administrative levels. A very interesting method to save transport cost was to pay the troops silver instead of copper cash or to give them tea or silk as a means of payment for the additional food.[83] The markets in the Western Territories willingly accepted this method of payment instead of cash, all the more as the Chinese cash coins had not yet been introduced in all communities. The Muslim towns in the Western Territories had their own currency, the *pul* coin that was only given up and replaced by copper cash after the conquest of the region by the Qing. The Dzungar campaigns were an important step towards the bureaucratization and standardization of war logistics. It was also seen that the responsible conduct and farsightedness of the provincial governments was a crucial element in the cooperation of the military staff with the civilian administration. Peter Perdue adds that the sufferings of the province of Gansu during the many campaigns in the west were balanced by a better integration of this remote region into the empire. The salaries of the many troops, furthermore, contributed to an economical regeneration

80 Lai, *Qianlong zhongyao zhanzheng*, p. 225.

81 In the Russian campaign of 1812, for instance, half of the horses died during the first few weeks of the march on Moscow. Van Creveld, *Supplying War*, p. 66.

82 Perdue, "Qing State and Gansu", p. 111.

83 Lai, *Qianlong zhongyao zhanzheng*, p. 226.

of the local markets.[84] This transformation of the role of Gansu is similar to the transition of the province of Sichuan from an economically and strategically unimportant region to a base area for a number of military campaigns.[85]

The Forgotten Issue of Maritime Transport in the Taiwan Campaign

The limits of the bureaucratization of warfare could be seen during the Taiwan campaign in 1786, when 40,000 troops from Fujian, Hunan, Guizhou, Guangdong, Sichuan and Zhejiang were shipped across the Taiwan Strait.[86] The war-waging province of Fujian organized several ports where the troops embarked (mainly Xiamen, Hanjiang near Quanzhou and Minxian near Fuzhou) and disembarked (Danshui near modern Xinzhu [not the modern Danshui], Luergang near modern Tainan [back then 'Taiwan'] and Luzaigang near Zhanghua). The logistic challenges were very different from and in some fields more complex than during the western campaigns or wars in mountain territory. Half of the troops were new recruits that had first to be equipped with uniforms and weapons. The 2,000 troops from Guizhou, 3,000 troops from Guangdong and 1,000 Banner troops from Guangdong travelled almost the whole way across several provinces along rivers, mostly downstream, so that the march to the war theatre lasted one month, while the way back upstream required three months. The 2,000 troops from Hunan used a similar method of travelling, just like the 2,000 troops from Sichuan that travelled the Yangtse River downstream and then entered the province of Fujian from Jiangxi. These troops were only partially Green Standard troops. Around 1,600 of them were recruits from the military agro-colonies around Maogong, the former Jinchuan, and 400 were native auxiliaries. It was believed that the inhabitants of the former Jinchuan were well suited for the hilly regions of central Taiwan, yet it was not borne in mind that the climate in Eastern Tibet was very different from that in the coastal plain in Taiwan, where subtropical heat and diseases exhausted persons not accustomed to them. Physicians therefore prepared special drugs against febrile diseases.[87] Furthermore, the troops from Maogong neither possessed uniforms nor weapons. It was therefore decided to have the necessary funds contributed by

[84] Perdue, "Qing State and Gansu", p. 112.
[85] Dai, *Sichuan Frontier and Tibet*, p. 236.
[86] Lai, *Qianlong zhongyao zhanzheng*, p. 131.
[87] Lai, *Qianlong zhongyao zhanzheng*, p. 133.

local officials from Sichuan.[88] Another problem was that the natives from Maogong had to be supplied with a special kind of diet, namely barley flour instead of rice, and with a lot of dried meat. Finally, many natives from Maogong did not understand Chinese, which made it necessary to take interpreters along. There were also 3,000 Green Standard troops from Zhejiang, belonging to a 'marines' garrison, and 1,000 Banner troops from the garrison in Zhapu who were directly shipped to Fujian before being ferried across the sea. Unlike during the other wars, there was only a very small number of high officials from the Banner garrison in Beijing that travelled to Fujian.

The ships to be organized to bring the troops across the sea had also to transport their supplies of food, bullets and gunpowder, and quite presumably also a minimum number of horses. There were about 100 ships organized to take all the troops from the ports in Fujian across the sea, or 120 ships for about 5,000 troops.[89] While land transport was seldom affected by weather, not all contingents were so lucky as to have a smooth crossing. Some ships, driven off course by wind and waves, arrived somewhat later at their destination. Almost half of the troops had never seen the sea before, a circumstance that the emperor respected by rewarding the troops from the inland provinces a present of 2 *liang* for each soldier.[90] In Luzaigang, a temple of the 'Heavenly Queen' Mazu was erected as the first official temple in Taiwan.

The Taiwan campaign was the first war in which the War Expenditure Code was applied. In the same instance, the campaign proved that there were several issues not yet covered by concrete regulations, like the payment for sea transport. The *Junxu zeli* only included regulations for the cost of transporting troops along inland waterways, with a defined landing phase during the night (corresponding to the logistics station). This is all the more surprising as there had been one instance in the past when troops from the mainland had to be shipped across the Taiwan Strait, viz. during the war against Zheng Chenggong (Koxinga) and his dynasty, yet this was a century earlier. The rebellion in 1732 of Lin Wuli and Wu Fusheng had been suppressed with the help of local troops, and only a small number of soldiers from the mainland had crossed the sea during that incident. The sectarian uprising of Chen Zongbao and Zheng Chun

88 Lai, *Qianlong zhongyao zhanzheng*, p. 129.
89 Lai, *Qianlong zhongyao zhanzheng*, p. 135.
90 Lai, *Qianlong zhongyao zhanzheng*, p. 135.

in 1770 could also be brought down by the prefect (*zhifu*) of Taiwan, Zou Yinyuan, with the help of local troops. There were apparently no precedent regulations for the shipping of such numerous troops as needed in 1786, nor about expenditure for ferrying an army across the sea. It was therefore decided to practice a payment according to the size of ships, as was the custom in brown water transport. For the concrete practice in the campaign it proved useful to count the distance from Xiamen to Luergang as 12 stations, the distance between Hanjiang and Luzaigang as 9, and the way from Min'an to Danshui as 8 stations.[91] Supplies, especially rice, was accounted for at a transport price of 0.036 *liang* per *dan* and station, which is in conformity with that of downstream transport on a river.[92] Other problems that could not sufficiently be resolved were the daily rations for the seamen, the price and amount of drinking water to be bought for sea passages, and the peculiar situation of the province of Fujian, where it was nearly impossible to recruit locals for labour according to the household registers. Instead, the local officials had to hire foreigners (*youke*) to carry the baggage of the troops marching from Jiangxi through the interior of Fujian.[93]

The food supplies sent to Taiwan did not only serve to feed the troops but also the local population that suffered from the martial activities of Lin Shuangwen and the imperial troops fighting his rebels. At first sight this seems much like support for war refugees, making the paternalistic Qing government appear as a quite modern regime. Yet it is important to know that the main objective was to impede the rebels from recruiting fighters among the local population. It was therefore common to train the able-bodied males of the villages in Taiwan as 'righteous people' (*yimin*) or 'village braves' (*xiangyong*), in order to create village militia for self-defence, while those parts of the population not able to defend themselves were assembled in strategic places and fed by the government.[94] During the Taiwan campaign 47,900 troops of village militia were recruited and paid

91 Lai, *Qianlong zhongyao zhanzheng*, p. 135.

92 *Hubu junxu zeli*, juan 5, fol. 2a–2b.

93 Lai, *Qianlong zhongyao zhanzheng*, p. 136.

94 Village militiamen were already during the Song period (960–1279) perceived as an ideal instrument not only of responding to heightening banditry, but also of cutting military expenditure. Paul Jakov Smith, "Shen-tsung's Reign and the New Policies of Wang An-shih", in *The Cambridge History of China*, vol. 5, *Part One: The Sung Dynasty and Its Precursors, 907–1279*, ed. by Denis Twitchett and Paul Jakov Smith (Cambridge: Cambridge University Press, 2009), p. 407. For a short moment I was trapped into an association with Ngo Dinh Diem's 'strategic hamlets'.

according to the statutes of regular troops, while almost half a million of refugees were supplied with food.[95] The rice production of Taiwan was actually relatively good, and a lot of rice was normally sent to the mainland, yet for the war, rice had to be imported not only from Fujian but also from other far away provinces, to an amount of more than a million *dan* of rice, to which more than 100,000 *dan* (10,300 m^3) of sweet potatoes have to be added as a staple food in Fujian and Taiwan.[96] The province of Fujian itself was able to deliver 3,400 *dan* (350 m^3) of rice from district granaries, a further 400 *dan* (41 m^3) came from various smaller districts, as well as 125,000 *dan* (13,000 m^3) of husked rice as rations for the troops and the labourers, taken along with them, and 13,700 *dan* (1,400 m^3) of rice produced in Taiwan. The province of Sichuan delivered 500,000 *dan* (51,800 m^3) of rice. It was far away but had been favoured with excellent harvests in the years before. What is more, all the rice could be shipped downriver along the Yangtse to Shanghai, and from there on to Quanzhou. By this method the cost per *dan* was no more than 1.56 to 1.67 *liang*. The province of Jiangxi contributed 150,000 *dan* (15,500 m^3) of rice that was shipped across the mountain range between Jiangxi and Fujian, also along smaller and larger rivers, partially on rafts, and partially by boat. 160,000 *dan* (16,600 m^3) of rice came from the province of Zhejiang. All this rice originated in district granaries. Smaller amounts were bought on the market, for instance in Danshui, where 4,200 *dan* (435 m^3) came from, or Luzaigang, where 1,400 *dan* (145 m^3) of rice were bought, at a price of more than 2 *liang* per *dan*. The Banner troops from Guangdong brought 1,200 *liang* of their own rice.[97] The ships transporting the rice across the sea were often prone to pirate attacks. The responsible supervisors of the logistics lines had to compensate the government for losses thus incurred. At least they did not have to pay for losses which occurred when ships, freight and men perished in the sea, especially in autumn, when the region was beset by typhoons. For those drowned the government paid indemnities, as provided for in the *Junxu zeli*.[98] The supply of rice from the mainland and from other provinces was so excellently organized that almost a third of the procured rice was left over after the termination of the campaign. It seems that for the Taiwan campaign of 1786, the whole empire cooperated in a well-concerted manner and brought the war to

---

95 Lai, *Qianlong zhongyao zhanzheng*, pp. 251–252.
96 Lai, *Qianlong zhongyao zhanzheng*, p. 248.
97 Lai, *Qianlong zhongyao zhanzheng*, pp. 248–249.
98 *Bingbu junxu zeli*, juan 4, fol. 7b; juan 5, fol. 4a.

such a smooth end that in this case one can almost speak of a perfect bureaucratization of warfare.

War Logistics During the Second Jinchuan Campaign

The second Jinchuan campaign was the largest, longest and most expensive war that the Qianlong emperor led. It was fought under highly unfavourable topographical conditions, in barely accessible mountain valleys. The troops had to be supplied along steep footpaths crossing mountain passes that were hard to manage during the winter months. Many thousands of labourers were therefore needed to take supplies to the camps and to the logistics stations themselves because the recruited labourers had to be fed, too. The long war of attrition against the easily defendable mountain fortresses of the natives also required a huge amount of military equipment, especially gunpowder, and adequate equipment to give the troops shelter and rest in the unfriendly climate. Horses, on the other hand, were a minor problem, at least in the war theatre itself, because they were of no use anyway, neither for transport nor for cavalry charges. The unusual conditions also produced a further phenomenon that was not seen during other campaigns, namely the support of the official transport system by private merchants commissioned to transport rice to the camps. This sub-chapter will explain in detail how the logistics issue was organized during the second Jinchuan campaign in order to demonstrate how much the Qing government had learned from past experience and therefore successfully brought together various actors and institutions of several administrative levels. They orchestrated all mechanisms necessary to ensure the smooth supply of a huge amount of persons over a long period of time.

The March to the War Zone and the Triumphant Return

Of the 144,000 troops participating in the second Jinchuan campaign 13,000 were Bannermen from Chengdu (Sichuan), Yongchang (Yunnan; actually troops from Beijing), Jingzhou (Hubei), Xi'an (Shaanxi), Beijing and from garrisons in the northeastern provinces of Heilongjiang, Jilin and Shengjing (modern Liaoning), 101,000 Green Standard troops from Sichuan, Guizhou, Shaanxi, Gansu, Hubei, Hunan and Yunnan, as well as 30,000 native auxiliary troops provided by eighteen local kings.[99] With

[99] *Jinchuan junxu li'an*, juan 1, fol. 7a–15b.

view to logistics the native auxiliary troops can be neglected because their home villages were not very far from the war theatre. Another battalion of logistically minor importance was the small contingent of 263 Banner troops that had stayed in Yongchang, Yunnan, as rear guard of the last Myanmar campaign and was brought again into the field at the beginning of the second Jinchuan war. These Banner troops were experienced soldiers, but from the logistics side their march was a rather decent yet well-executed matter of organization. A report of this march by a participant has survived in Wang Chang's small book *Shujiao jiwen* "Account from the borderlands of Shu (i.e. Sichuan)", where he describes which parts of the way were covered on foot or riding, and which by boat. The Banner troops from Yongchang, Yunnan were accompanied by 800 horses and 200 mules.[100]

The numbers of troops recorded in the logistics and expenditure statutes *Pingding Liang Jinchuan junxu li'an* are consistently divided into contingents that were dispatched to Jinchuan before the disastrous defeat at Mugom in summer 1773, and those contingents that were part of the massive troop reinforcement after that date. A different group of soldiers were those sent to put down the unruly Jinchuan natives at the beginning of the campaign. The men were still seen as being within the framework of normal 'constabulary actions' and not as troops deployed for a war, and were therefore registered in the files for 'barbarian affairs' (*yiwu'an*), and not in those for war expenditure (*junxu'an*). This is unfortunately the only chronological segmentation in this document, so that the concrete size of the contingents and the dates when they marched from their home garrisons to the war theatre can only crudely be documented with the help of the official annals of the campaign, the *Pingding Liang Jinchuan fanglüe*, the edict collection "Veritable records" *Qingshilu*, or scattered documents from the State Council (*Junjichu dang*) or the Palace Archives (*Gongzhong dang*).

Between May and September 1771 governor-general A'rtai dispatched 6,190 native auxiliary troops and 8,415 Green Standard troops from Sichuan that were supported by 3,960 so-called long-distance porters (*changfu*), probably supplementary troops or manservants. In October 1771, the new acting governor-general Defu and general Wenfu brought in the first contingent of elite Banner troops from Yunnan (between 263 and 280 men) mentioned above. The next detachments were already to be found among

100 *Pingding Liang Jinchuan fanglüe*, juan 13, fol. 26a–26b (QL 36/12/gengyin).

the files for war expenditure (*junxu'an*), 18,421 men in the old files (*jiu'an*, including the 8,415 men from summer 1771) being compiled before August 1773, and 6,906 in the new files (*xin'an*) being recorded after that date. In total, the province of Sichuan dispatched 25,267 (or 25,327) Green Standard soldiers. The next contingent was the Banner troops from Chengdu. In two batches 1,444 Chengdu Banner troops were dispatched before August 1773 and 318 men later. This makes a total of 1,762. Of greatest importance were the elite Banner troops from the northeastern provinces. The old files include only 280 elite Banner troops, while the major part of the 6,497 men are included in the new files, after the Mugom catastrophe, when the emperor had dispatched elite troops from the Firearms Brigade (*huoqiying*), the Scouting Brigade (*jianruiying*), and troops from Jilin and Heilongjiang which were believed to be the only reliable means to put down the ungovernable Jinchuan people. The important Banner garrison of Jingzhou in the province of Hubei supplied 2,202 troops who were all dispatched after the summer of 1773. A last important Banner garrison was Xi'an in Shaanxi, which provided 2,041 troops dispatched after the Mugom disaster. The figures listed in the *Jinchuan junxu li'an* confront the reader with the problem of 1,000 people missing. The documents in the *Pingding Liang Jinchuan fanglüe* nevertheless give evidence that in fact 2,000 troops were dispatched from Xi'an to Jinchuan.[101]

By far the largest part of the imperial troops engaged in Jinchuan were Green Standard troops. Although almost a quarter of them came from the province of Sichuan, large contingents were dispatched from neighbouring provinces. 15,388 troops came from the province of Guizhou, 5,118 of whom were dispatched before August 1773, and 10,270 after that date. From various Green Standard garrisons in the two provinces of Shaanxi and Gansu, 36,643 soldiers, and several thousand supplementary soldiers (*yubing*, *yuding*) were dispatched. 22,493 regulars and 7,500 supplementary troops had been deployed before August 18, 1773, and 6,650 regular troops after that date. Hubei and Hunan provided 15,940 troops, out of which number 5,091 regular troops and 1,700 supplementary troops were dispatched before August 1773, and 8,149 regular and 1,000 supplementary troops after that date. This makes for a total of 2,700 supplementary troops and 13,240 regular troops.[102] From Yunnan province finally, 8,097 Green

[101] *Jinchuan junxu li'an*, juan 1, fol. 11b. *Pingding Liang Jinchuan fanglüe*, juan 65, fol. 24a (QL 38/7/dingyou); juan 68, fol. 8a (QL 38/7/renzi); juan 71, fol. 7a–7b (QL 38/8/gengwu).

[102] There were also supplementary soldiers from Sichuan, yet exact numbers are not known. The ratio, according to the precedents, was 30 supplementary soldiers for

Standard troops were dispatched, 3,036 being listed in the pre-Mugom files, and 5,061 in the new files.

Beside the Banner troops and the Green Standard soldiers, also many native auxiliary troops were dispatched by various native kings. Their number was as high as 29,597 and included officers (*tubian*), soldiers (*tubing*), and recruits from the military colonies (*tunlian*) in native territories recently incorporated into the empire. 6,100 of these troops were included in the 'barbarian affairs' files, 20,348 in the old files, and 3,149 in the new, post-Mugom files. The native kings of Litang, Mingdjeng, Batang, Wasi, Wori (Okshi), Djoktsai, Sunggang, Somo, Muping, Damba, San Aba, Chosgyab, Gebshidza, Hor-Djanggu, Derget, Rumi-Djaba, Badi, Bawang, Hannyu and some small countries in the sub-prefecture of Songpan dispatched auxiliary troops.[103]

Concerning the Banner troops from the northeast is has to be assumed that more than 1,000 came from various garrisons in the province of Jilin, a further 1,000 were provided from among the Solun (Ewenke) banners in Heilongjiang,[104] general Shuhede led his Oirat (Dzungar) contingent of about 1,000 men from Ili to Jinchuan,[105] and 1,000 other troops must have come from other garrisons in Heilongjiang province, under the command of Fengsheng'e.[106] Although it is known that Dzungar troops took part in the campaign,[107] the *Pingding Liang Jinchuan junxu li'an* does not list any particular number of Oirat troops from Ili.

---

100 standard troops. *Hubu junxu zeli*, juan 3, fol. 4a. *Jinchuan junxu li'an*, juan 2, fol. 28b–29a. In practice, four cavalrymen and five infantrymen were supported by one supplementary soldier (25 resp. 20 per cent of the standard troops), according to the precedents of the western campaigns. This would mean that of the 25,000 Green Standard troops from Sichuan, 8,833 were supplementary soldiers (alternatively manservants, *genyi*).

103 The list includes more than 18 allied kingdoms. On the other hand, the important country of Dzagu is missing in this list, although its king dispatched several thousand soldiers (see for example *Pingding Liang Jinchuan fanglüe*, juan 5, fol. 19a [QL 35/8/gengchen]; juan 12, fol. 19a [QL 36/12/bingzi]; juan 49, fol. 23a [QL 38/1/yichou]).

104 The Solun unit consisted of 2,000 men (Ewenke, Daghur, and Oročen), which means that it could also have been dispatched as a whole. But as the emperor wanted to deploy elite troops with enough firepower, musketeer contingents from elsewhere in the province could also have been dispatched. Compare *Baqi tongzhi*, juan 35, pp. 615–617.

105 *Pingding Liang Jinchuan fanglüe*, juan 41, fol. 7a (QL 37/10/dinghai); juan 66, fol. 3b (QL 38/7/gengzi). Lai, *Qianlong zhongyao zhanzheng*, 116–125, does not even mention the Oirat troops in his specification of the march routes through the provinces, although he indicates in his map (Map 1) that there were indeed troops coming from Ili.

106 *Pingding Liang Jinchuan fanglüe*, juan 63, fol. 13b (QL 38/6/xinhai).

107 For example, *Pingding Liang Jinchuan fanglüe*, juan 38, fol. 16a–16b (QL 37/9/dingwei), naming De'rsenbao, Pu'rpu and Yulusi. *Idem*, juan 41, fol. 7a (QL 37/10/dinghai) names Shuhede, who was general of Ili. *Idem*, juan 43, fol. 6b (QL 37/11/yimao). Shuhede

It was planned to dispatch 2,000 more troops from the Capital at an earlier point of time, in QL 37/2 (Mar 1772),[108] but these were finally replaced by 5,000 Green Standard troops from Shaanxi and Gansu, because the garrison in Beijing was too far away from the war theatre as that the elite troops could be brought into operation, and because Banner troops were much costlier than Green Standard troops.[109]

Concerning the figures for the Green Standard troops, it becomes evident that Sichuan itself provided about a quarter of all Green Standard troops, while the most part of the Green Standard units came from neighbouring provinces. Yunnan, whose troops had taken part in the Myanmar campaigns, did only procure a smaller number of troops, while the main part came from the area of Shaanxi and Gansu, the provinces that had regularly provided the fighting units for the wars in the Western Territories and that were, at the end of the Qianlong reign, heavily reinforced as part of the emperor's defence policy.[110] After the territories in the west and Tibet had been largely pacified, the units of those two provinces could be freely used for the Jinchuan campaign. The rest of the troops came from the neighbouring provinces of Hubei, Hunan and Guizhou, whose troops did not only have to march a distance, which was not too long, but had also experience in fighting insurgent native tribes like the Miao in Guizhou. Troops of those provinces were all experienced combat units and were permanently deployed for the numerous wars which the Qing emperors waged. It was rarely seen that troops from eastern provinces like Jiangxi, Jiangsu, or Zhejiang were used for campaigns—except, of course, the invasions of the island of Taiwan. The heartland of the empire should not be deprived of its troops.

The march of large contingents of Banner troops from Beijing and the northeast required a huge amount of means of transport. The troops marched in batches of 200 or 300 troops, which was a smaller size than during the western campaigns, in which the average size of batches had been 500 soldiers. The troops from the northeast took a short rest in Beijing

---

was ordered to choose 1,000 brave Oirat troops to be sent to Sichuan under the command of Chengguo. *Idem*, juan 63, fol. 15b (QL 38/6/xinhai).

108 *Gaozong shilu*, juan 902, fol. 9b–11a (QL 37/2/dingmao). Lai, *Qianlong zhongyao zhanzheng*, p. 116. It was pre-calculated that the cost of deploying 2,000 Capital Banner troops would be as expensive as sending 10,000 Green Standard troops from a province located closer to the war-waging province Sichuan.

109 *Gaozong shilu*, juan 903, fol. 18b–19a (QL 37/2/dinghai); juan 909, fol. 14a–14b (QL 37/5/wuwu).

110 Dai, "Yingyun shengxi", p. 52.

and then marched on in 12 contingents of 300 men each. The Beijing units marched in 12 contingents of 200 men.[111] This was necessary because the Banner troops from Beijing were allowed to use a larger amount of horses and manservants. The prefects and district magistrates of the province of Zhili (modern Hebei) therefore had to assemble more than 2,000 carts in the province. A large part of these had to be organized from villages far away from the large courier stations where the troops marched through. Riding horses and beans to feed them were also to be procured by the civilian government in Zhili. The troops needed eight days to march from Beijing to the border of Henan, and from there eleven more days until they reached Shaanxi. Each of the stations in Henan was equipped with 250 carts, which made a total of 2,750 vehicles for the whole province. For crossing the Yellow River near Zhengzhou, 30 to 45 boats were hired that ferried the contingents to the other side. The provincial administration commissioner of Shaanxi, Bi Yuan, bought sufficient fodder for the horses in advance, and governor Le'rjin procured an adequate amount of carts that were not allowed to reach the assigned logistics station earlier than five days before the troops arrived. Both officials cooperated with each other because in autumn the roads of Shaanxi were quite muddy and nearly unmanageable for carts. Bi Yuan therefore organized large carts drawn by a team of four mules or donkeys, as well as pack saddles for mules to transport the baggage. In difficult terrain the baggage would be reloaded from the cart on the backs of the mules. For the critical passage of the plank road viaducts in the Qinling Range between Shaanxi and Sichuan Le'rjin ordered 2,000 transport animals from the province of Gansu. The bad experience made in the province of Sichuan during the first Jinchuan war had a very positive effect on the logistics organization in the present campaign. Fulehun, governor-general of Sichuan, provided a sufficient number of porters in the logistics stations and asked Bi Yuan to lend 4,000 mules for the transport from the provincial border to Chengdu.[112] Yet this was not the only case, as a document from a few days later shows that Shaanxi had already sent 7–8,000 mules to Jinchuan. A further lot of 4,000 (or 5,000) mules was bought in Shaanxi in the summer of 1773.[113] Such transactions were sometimes even handled spontaneously.

---

[111] Lai, *Qianlong zhongyao zhanzheng*, p. 117.

[112] *Pingding Liang Jinchuan fanglüe*, juan 37, fol. 13b (QL 37/8/gengchen), 17a–17b (QL 37/8/wuzi).

[113] *Pingding Liang Jinchuan fanglüe*, juan 67, fol. 5b (QL 38/7/bingwu); juan 70, fol. 9b–12b (QL 38/7/xinyou).

Thus the same document declares that 1,500 pack horses from Hubei were to stay in Chengdu rather than being sent back. Both porters and mules were used for long-distance transport covering a distance of two days. This method helped to avoid organizational bottlenecks because the stations were assigned a sufficient amount of transport means. From Chengdu to the war theatre riding horses were provided for all troops inside provincial territory, yet from the border on to the native kingdoms, only officers were allowed to ride, while common troops were paid a compensation. How difficult it was to move in the mountains of Jinchuan can be seen from the speed of the troops that needed six weeks to march from Beijing to Chengdu (1,800 km by road), yet 20 or 30 days to cover the distance from Chengdu to the camps (270 km to Lesser Jinchuan and 420 km to Greater Jinchuan). Western troops used to march with a normal speed of 15 miles (about 25 km) per day for infantry including the baggage train, while artillery units were only half as quick.[114]

The troops from Ili in the far west and from Gansu crossed the Qinghai Plateau and liaised with the fighting corps on the northernmost part of the front. The Green Standard soldiers of Hubei and Hunan used to march in contingents of 500. Governor Chen Huizu had organized 3,000 horses for the Banner troops of Jingzhou, which was more than sufficient, and yet the Banner soldiers offered marching by foot, and were rewarded with one monthly salary for their heroism. The emperor also rewarded troops from Guizhou that had marched the long way to Jinchuan by foot.

It was only in the western parts of the war theatre in Jinchuan that cavalry attacks were feasible. There was at least one incident in which the Jinchuan enemy staged a cavalry attack on Qing units on August 25, 1774. The emperor therefore immediately ordered to provide 180 garrison horses to support the troops against hostile charges. Yet the maintenance of those horses proved to be so expensive (with daily cost of 0.8 *liang* per animal) that it was soon decided to give up war horses.[115] There are several statements to be found in the documents which prove that the inhabitants of the native kingdoms of the west, like Derget, as well as the Jinchuan rebels, were in possession of horses and also rode horses during attacks.[116] Some of the natives more to the west were even more accustomed to riding than to walking, and therefore were of no great use

---

114 Van Creveld, *Supplying War*, pp. 2, 12.

115 According to Lai, *Qianlong zhongyao zhanzheng*, p. 331.

116 *Pingding Liang Jinchuan fanglüe*, juan 59, fol. 5b–6a (QL 38/4/jiachen); juan 64, fol. 12a–12b (QL 38/6/bingchen); juan 100, fol. 7b–8a (QL 39/7/yiwei).

when staging non-mounted attacks.[117] Horses were made use of by the imperial troops when trying to capture fleeing enemy chieftains. For this purpose the generals had to buy horses for the Banner and Solun troops who had the task to catch all the enemy leaders.[118]

The empire-wide valid *Junxu zeli* does not prescribe any particular regulations for the triumphant return (*kaixuan*) of the troops. Contrary to this situation, the Jinchuan regulations contain a lot of particular rules for the troops returning home, beginning with the luggage for each officer returning home by waterway, to the horses which were to be provided to the troops when returning home or hastening to the Capital.[119] The reason for regulations was not that the triumphant return played an important part for the troops (which it indeed did for the government when the Capital troops brought home the war-prisoners and the Banner units paraded before the emperor), but that the troops would travel back to their home garrisons as fast and inexpensively as possible. Only upon arrival of troops in their home garrison the government ceased to pay the expensive salt-and-vegetable money. The Ministry of Revenue was certainly not willing to spend more than really necessary on the transport horses of the victorious troops. It was even prescribed that an officer was not to take with him more than 100 *jin* (about 60 kg) of luggage, a soldier a maximum of 25 *jin* (about 15 kg) when travelling by boat. While the Ministry of Revenue was willing to pay the transport fee for these loads, the owner had to bear the transport cost for excess luggage.[120]

The return of the troops was already prepared as early as autumn 1775, when the largest castle of the lord-lama of Jinchuan had been conquered.[121] The Banner troops from Beijing and the northeastern provinces were to travel back by boat, which was the most comfortable and also the quickest method of travelling. They travelled down the Yangtse River and then marched through Xiangyang (modern Xiangfan) across northern Hubei, Henan and Zhili to Beijing. For this march the local governments of these provinces had to organize 18,000 carts in total, as well as 80 boats for crossing the Yellow River. For the festivities of the triumphal return at the dynastic altars, 1,600 troops were selected to participate, and were to

117 *Pingding Liang Jinchuan fanglüe*, juan 47, fol. 13a–14a (QL 38/1/jiawu).

118 *Pingding Liang Jinchuan fanglüe*, juan 100, fol. 8a–8b (QL 39/7/yiwei); juan 120, fol. 22a (QL 40/7/dingsi).

119 *Jinchuan junxu li'an*, juan 2, fol. 34b–36a, 72a–72b.

120 *Jinchuan junxu li'an*, juan 2, fol. 35b, 72b.

121 *Jinchuan dang* 40/III/00273 (no date).

be shipped to Beijing as fast as possible.[122] These troops marched from the camps to Chengdu, and on via the provinces of Shaanxi, Henan and Zhili. Outside provincial territory, the logistics stations for the campaign could be used, yet they were equipped with the smallest number of porters possible. Green Standard troops were paid salted-vegetable pay and were given daily rations on their way from the camps to Chengdu. After passing the border to Sichuan province, both were halved, except for the manservants. Higher officers were not given any rations, officers of the rank of company commander (*qianzong*) and lower received a daily pay of 0.01 *liang*. Outside provincial territory, which means, still within the mountains, a daily distance of 2 to 3 stations had to be covered. Officers and manservants with no horses or porters at their disposal could hire a porter at 5 *fen* a day. Riding horses were likewise not provided to the common troops. For the stretch from Chengdu to the border of Shaanxi, 3,500 horses served to transport the troops and their baggage. The province of Shaanxi had to lend out 2,000 horses from Gansu and 1,000 horses from Shanxi because the garrisons in the province were virtually depleted of quadrupeds. At least, the private market provided a number of 700 horses hired for the transport of the troops back home.[123] The troops also marched along the shorter route through the province of Shanxi because in this way the expenditure for two stations could be saved, and the province had still some reserves in manpower and transport tools because it had not provided troops to the campaign and not participated in the organization of the march to the battlefield.[124]

Troops from the Capital and those participating in the official welcome were given tallies (*kanhe*) to be presented at the courier stations and were paid 0.1 or 0.2 *qian* per day and 0.05 *qian* for each manservant. This pay was to be advanced from the Sichuan government for the way to the border of Sichuan (eleven days). The horses for the troops had to be prepared by the courier stations, and only if this source was depleted, private horses were hired. The salted-vegetable pay for the Banner troops likewise had to be advanced by the province of Sichuan. Troops having pack-animals but no riding horses at their disposal were to use the pack-horses for riding and were given porters to transport their luggage, the latter being recruited from the logistics stations. Only if no station porters

122 Lai, *Qianlong zhongyao zhanzheng*, p. 117.
123 *Jinchuan dang* 40/IV/00317–321 (QL 40/12/20).
124 *Jinchuan dang* 41/I/00053 (QL 41/1/11), 00075–76 (QL 41/1/11), 00100 (QL 41/1/17).

were available new people were hired from among the population. The few Banner troops staying in the Jinchuan area and only returning later were not allowed to be paid any additional porters or horses and were not given salted-vegetable pay (as the war was over then), but had to return to their home garrison as fast as possible. Each officer was only allowed one riding-horse.

For the Chengdu Banner troops, the salted-vegetable pay for ten days of travel had to be advanced by the camp the troops were leaving. Each day two smaller stations (*xiaozhan*) were to be covered, and each person was given a ration of 1 *sheng* of rice daily. Only troops with heavy equipment like cannons were allowed to hire additional porters.

From Baxian (Chongqing) on the troops from Hubei and Hunan, as well as those from the Capital and the northeast, travelled by boat. The troops of Guizhou and Yunnan were assisted by 50 porters for every 100 men, and only officers of the rank assistant brigade commander (*shoubei*) and higher were not given salted-vegetable pay. From Chengdu, those troops marched to Leshan, from where the troops travelled by boat. Each person could hire a transport animal—porters were not allowed. The troops of these two provinces travelled exclusively overland, the Shaanxi troops via Guangyuan at the border between Sichuan and Shaanxi, and the Gansu troops via the Chaimen Pass in the prefecture Long'an.[125]

The regulations for water travel were similar to those for overland travel: both had to be carried out as fast as possible, for which purpose tallies were given which marked these troops as fast-travellers. From Chengdu to Baxian, ten riverine stations were called at, from Baxian to the border of Sichuan to Hubei, seven more stations. For each of those two stages the necessary money had to be provided, first in Chengdu, and then in Baxian. Officers of rank 6 and higher were allowed to rent their own boat,[126] officers of rank lower than 6 had to share one boat for two people. Soldiers and manservants travelled with 7 to 8 persons per boat.

The main means to ensure a speedy travel during the triumphant return were to hand them over tallies allowing the troops to be served faster in the courier stations, rations in advance for the period of time until they reached the border to Shaanxi, and halving the salted-vegetable pay which automatically force the troops to speed up.

125 *Jinchuan junxu li'an*, juan 2, fol. 34a–37b, 137a.

126 State offices, civilian and military, were graded into 14 half-ranks, 1a being the highest, and 7 and 8 the lowest categories. Rank 6 corresponds to a department vice magistrate or sub-prefectural magistrate, or to a company commander.

The troops from Yunnan and Guizhou would march to Yazhou where they boarded the boats for their return, as described above. The other 50,000 troops were to pass the western and northern routes. For the provisions outside provincial territory, considerable quantities of rice had to be stored in the grain stations: 500 to 600 *dan* (52–62 m^3) per station was estimated as sufficient.[127]

Feeding the more than 10,000 Gansu troops marching through the barren northwestern prefectures of Sichuan would be quite expensive. Therefore it was—as far as possible—ordered that the grain stored in Maozhou (see Map 2, p. 195) should be used, amounting to 4,000 *dan* (414 m^3) of rice and 8,400 *dan* (869 m^3) of barley, as well as the 1,100 *dan* (114 m^3) of rice from Songpan.[128] The rations for the barren regions in Gansu were to be advanced by the province of Sichuan and the respective sums later to be paid back by Gansu.

When hiring mules it could be difficult when the beasts had to be rendered back to their owners. Each animal was therefore marked with a small wooden plate probably hanging from its neck, indicating the name of the soldier(s) whose equipment the beasts were bearing, and the name of the owner.[129] After delivering the silver from the Capital, being the fund for officials' salaries, 2,500 mules from Shaanxi were to be given back to their owners. There were also some 2,000 to 3,000 mules hired in various districts in the province of Sichuan because the harvest period had just begun, which would make it impossible to recruit porters from among the peasantry to carry rice to the camps.[130] In order to cover the long distances of the western branch of the southern route which led through the territory of the native kings of Mingdjeng and Chosgyab, the logistics officials hired local owners of mules to carry rice and equipment.

Three factors had decisively contributed to the smooth organization of the movement of troops over thousands of Chinese 'miles' and through many regions, some of which were barely inhabited. The first was an early planning of the march to the war theatre and back, which gave the organizers more time to schedule and arrange, especially for the notorious bottlenecks in the hills of Shaanxi and the river gorges of the Qinling Range. The second was a more relaxed forwarding of those troops that were not obliged to arrive in the Capital for the triumphal festivities

---

127 *Jinchuan dang* 40/IV/00151–152 (QL 40/10/29).
128 *Jinchuan dang* 40/IV/00249 (QL 40/12/3).
129 *Pingding Liang Jinchuan fanglüe*, juan 69, fol. 19a–19b (QL 38/7/gengshen).
130 *Pingding Liang Jinchuan fanglüe*, juan 105, fol. 8a–9b (QL 39/9/jiazi).

arranged by the emperor. Batches were smaller, and the owners of carts and beasts were allowed to be earlier ready to deliver their services at the stations. As a compensation for this loss of time and money, the batches were marching in intervals of two days, which means that the carts were shuttling between two stations permanently and without a day of repose, until, after twenty-four days of work, twelve batches of troops were successfully brought closer to the destination. The third factor was the effective implementation of water transport for those troops that were not forced to arrive in Beijing or the home garrisons at a fixed date. Although water transport was not easy to plan in respect of time, it was a relatively cheap and less laborious means of bringing troops back home. The troops urgently called to the Capital were given tallies giving them the same status as express couriers, and were given rations in advance, which saved the time to change money or purchase foodstuff.

Organization of the Logistics Lines Near the War Theatre

The arrangements of taking troops from their home garrisons to the war theatre and back were temporary, yet often based on a permanent system of courier stations throughout the empire. Outside the provinces, where no network of courier stations existed, the logistics stations were established on a thoroughly temporary basis. At least, the experience from the first Jinchuan war could be of advantage when building up logistics stations in the same place or in another location during the second Jinchuan war. The same operating experience facilitated the recruitment of labourers, either from the peasant population in Sichuan according to the tax registers, or from among the natives in Jinchuan and its surroundings.

The high mountains of the Qionglai Range complicated the access to the river valleys of Lesser Jinchuan and Greater Jinchuan. The logistics managers therefore made use of four logistics lines that gave access to the land of the enemy from different directions. The most direct route, the so-called 'western route' (*xilu*), more or less led westwards from Chengdu, via Guanxian (modern Dujiangyan), the 'Crouching Dragon' Pass Wolongguan, and the pass of Mt. Sigulang (Siguniang) down into the valley of Rilong, Okshi (Wori) and Lesser Jinchuan (modern Xiaojin). From Lesser Jinchuan the western route was extended northwards into the valley of the Lesser Jinchuan River and across the mountains in northwestern direction, to reach the valley of the Greater Jinchuan River. The 'northern route' (*beilu*) also started in Guanxian, passed Wenchuan and Dzagu (also called Dzagunao; modern Lixian) and reached the valley of

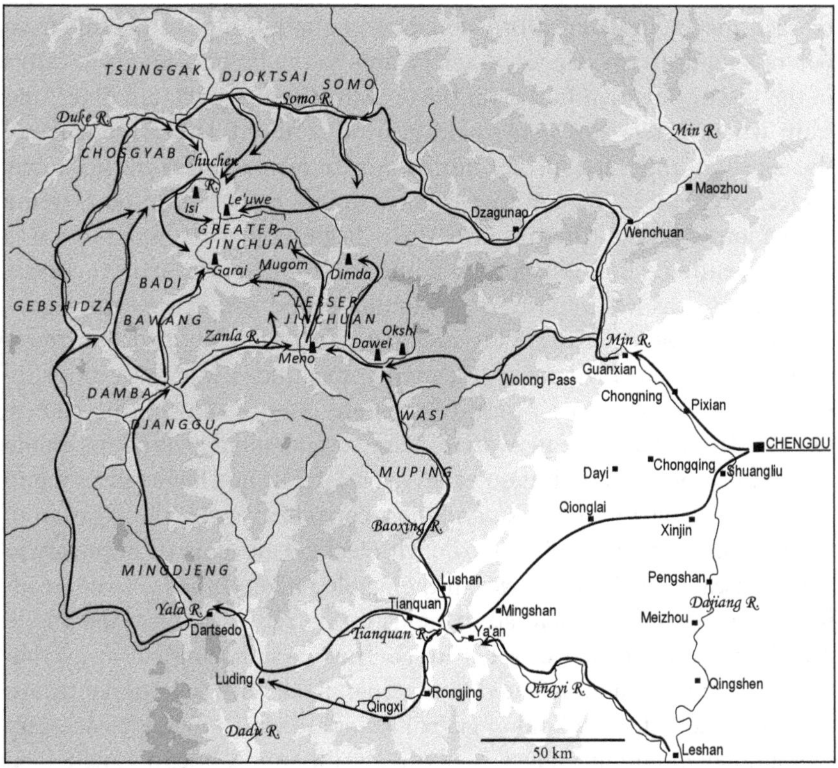

main prefectures, important border towns
N A T I V E K I N G D O M S
important native villages
▲ (war towers) *main castles and fortified monasteries*

Map created by the author.

Map 2. The logistics trails in the Jinchuan war theatre.

the Somo River, from which a network of logistics lines crossed the mountains southwards towards the heartland of Greater Jinchuan.

In 1774 a 'new western route' (*xin xilu*) was inaugurated that began near Dzagu and crossed the mountains westwards. It was harder to manage but proved more efficient than the other routes for the supply of the troops penetrating into Greater Jinchuan. The 'southern route' (*nanlu*) began in the hub port of Yazhou (modern Ya'an) southwest of Chengdu and crossed the Dadu River at the famous iron chain bridge at Luding, from where long-distance stages led northwards to the point where the waters of the Lesser Jinchuan River join those of the Greater Jinchuan River, and from where Lesser Jinchuan could also be attacked. Other branches of the southern route led further northwards to take supplies to the troops attacking Greater Jinchuan from the south and west, and in its extreme points joined the northern route. At the beginning of the war there was also a 'central route' (*zhonglu*) leading from Yazhou directly to Lesser Jinchuan. It was less important during the second half of the campaign.

It can be seen that these logistics routes directly supplied the troops attacking the two kingdoms of Jinchuan from all sides. Moreover, this supply system was not only a sum of separate lines leading to the camps, but consisted of a dense network of relay stations that covered the whole area, ran through all mountain valleys and left out no chance to feed and equip the troops in the fastest and most convenient way. The creation of such a dense network had also to do with tactical movements necessary to wipe out each single war tower in the whole region. It had proved insufficient to destroy the main strongholds of the rebel leaders—the whole area had to be cleaned of rebel nests. There was no unified army that could be attacked and defeated in the open field, but the imperial troops had rather to fight in a kind of guerrilla warfare which made it necessary to pound all mountain valleys. Accordingly, the supply lines also extended into all vales and up to all alpine meadows.

The organization of the supply lines was in the hands of the provincial administration commissioner (*buzhengsi*) of Sichuan, together with two to three circuit intendants (*daofu*) or prefects (*zhifu*).[131] High civilian officials were dispatched on imperial command (*qinchai*) and were appointed to certain locations or 'followed to the camps' (*sui ying*) to supervise a whole logistics route or a section of it or an important station like a camp. The list of the stations in the supply line in the *Jinchuan junxu li'an* mentions several times supervising civilian officials with the ranks of

[131] *Pingding Liang Jinchuan fanglüe*, juan 32, fol. 4a–4b (QL 37/6/jiashen).

prefects, circuit intendants, provincial surveillance commissioners (*niesi*), provincial administration commissioners (*fansi*), vice ministers (*shilang*), and governors-general (*zongdu*).[132]

Rank and task of the civilian officials sent to the war theatre managing the logistics and courier routes reflected their position in peace-time government: While single stations were managed by district magistrates, prefects had to oversee larger stations or a couple of smaller stations, and circuit intendants were responsible for whole routes. On the northern route, each grain station (*liangtai, taizhan*) was administered by one circuit intendant or prefect.[133] The participation in the military duty (*bingchai*) of the supply lines offered opportunities for promotion. The prefect (*zhifu*) of Ningyuan, for instance, had displayed greatest diligence in organizing the transport of rice along the transport routes through the land of Chos-gyab in the northwest and was therefore rewarded with the title of a circuit intendant.[134] If the transport did not go well, the responsible civilian official was removed from his office.[135] Just as in the top echelons of the commanders it was also common on the lower levels that a military official and a civilian official worked together to resolve transport problems, for example, one brigade commander (*youji*) and a prefect, the former commanding the troops protecting the logistics stations.[136] Some civilian officials were also appointed to supervise the camp foundries (*paoju*)[137] or had, as it was in their area of jurisdiction, to care for the supply of metal and the recruitment of casters.[138] Among those persons, by the way, many were waiting (*houbu*) for a position to fall vacant, which means that not all district magistrates serving in Jinchuan were actual heads of a district, but many virtually unemployed officials to be were used to fill the ranks of the organizers of logistics. By this management it was not only possible to test the usefulness of the candidates, but also to save a lot of money because officials on probation were not paid any salary.

Inside provincial territory the troops on their march to the battlefield served themselves from the horses in the courier stations, yet outside provincial territory, courier stations had first to be established and to be equipped with horses from the garrisons in the provinces. The Jinchuan

132 *Jinchuan junxu li'an*, juan 2, fol. 91a–122b.
133 *Pingding Liang Jinchuan fanglüe*, juan 78, fol. 17b–18a (QL 38/10/xinhai).
134 *Pingding Liang Jinchuan fanglüe*, juan 89, fol. 12a (QL 39/2/guisi).
135 *Pingding Liang Jinchuan fanglüe*, juan 103, fol. 22a–22b (QL 39/8/wushen).
136 *Pingding Liang Jinchuan fanglüe*, juan 65, fol. 6b–7a (QL 38/7/guisi).
137 *Pingding Liang Jinchuan fanglüe*, juan 123, fol. 19b (QL 40/8/jiyou).
138 *Pingding Liang Jinchuan fanglüe*, juan 101, fol. 23b (QL 39/7/yihai).

regulations contain a detailed list of the horses kept ready for the courier service.[139] Although the number of station horses (*zhanma*) on the particular transport routes was quite constant, there were great differences among the stations especially on the northern and the southern route. On the western route, the number of purchased horses in each station was between 20 and 30; in the eastern part of the route, which was opened at the beginning of the war, there were also kept ready between 4 and 20 horses from the garrisons. Chengdu provided 30 horses for the station within the city itself, the district of Xinjin sent 20 horses south to each station on the route to Luding, the district of Pixian sent horses north to the route to Wenchuan, 20 horses for each station.[140]

It was regulated that if there were not enough horses on the market, or if the purchased horses could not reach their destination in time, 1,500 garrison horses had to be dispatched and were distributed to all stations inside and outside the country. Before the horses reached the stations, runners had to transmit the mail. When military garrisons from Sichuan had to dispatch horses or soldiers to the war theatre the responsible official, in this case the provincial military commander (*tidu*), had to keep an eye on the problem so as not to totally evacuate all garrisons from troops and horses. At the beginning of the war, of the 2,300 horses in the garrisons therefore only 1,500 were selected for use in courier stations. Nor were the garrison horses useless, as they could be used for the courier service, which prevented squandering fodder for idle horses left behind in the then under-staffed garrisons.[141] That the numbers of horses in the particular courier stations showed considerable differences can also be seen from a document in the official history of the war, where 16, 20, 30, or 40 were standard figures for courier horses per station.[142]

At the end of each of the five route listings (western, southern, central, northern, and new western) in the Jinchuan war expenditure precedents, the total number of horses to be provided on the respective routes is indicated. Of the total of 4,187 horses on all routes,[143] 2,535 horses were purchased on the market and 1,500 borrowed from the garrisons. Lai Fushun

---

139 *Jinchuan junxu li'an*, juan 2, fol. 91a–122b.

140 *Jinchuan junxu li'an*, juan 2, fol. 83b.

141 *Pingding Liang Jinchuan fanglüe*, juan 36, fol. 14a–14b (QL 37/7/jiazi); juan 72, fol. 3b (QL 38/8/jimao).

142 *Pingding Liang Jinchuan fanglüe*, juan 81, fol. 13a–15a (QL 38/11/jiaxu).

143 *Jinchuan junxu li'an*, juan 2, fol. 95a, 102a, 111a, 120a, 122b. Lai, *Qianlong zhongyao zhanzheng*, p. 333, gives the total number of 3,294 horses for the courier stations, with 1,383 animals being provided by the garrisons, and the rest being purchased.

Districts
<u>Transport terminals</u>
(places otherwise mentioned in the text)
—— road transport

Source: *Jinchuan junxu li'an*, juan 2, fol. 123a–137a. Map created by the author, based on Tan, *Zhongguo lishi ditu ji*, pp. 39–40.

Map 3. Origin and transport of grain for military supply in Sichuan.

provides somewhat different figures: Before QL 37/6 (Jun 1772) 1,361 horses served in Jinchuan, about 850 of which were used inside and about 580 outside the country. After that date, 1,934 horses were required, so that the total number of horses used in the Jinchuan region increased to 3,295.[144]

For the transport of grain and military equipment, the province of Sichuan had provided 2,000 horses and mules, and more than 7,000 animals were hired from owners in Shaanxi, and from the Yongchang garrison in Yunnan some 1,000 horses and mules which had been left over from the Myanmar campaigns.[145] 2,000 other 'stout' horses came from garrisons in the province of Gansu, and 4,000 from Shaanxi, which were sent out to support the transport of military equipment along the plank road viaducts between the provinces of Shaanxi and Sichuan.[146] At the beginning of the campaign horses and mules had been collected from garrisons and had been bought on the market, "at a favourable price, and easy to get".[147]

Grain and Other Food

The *Jinchuan junxu li'an* contains a long list about the supply lines of grain within the province of Sichuan.[148] All districts which supplied rice to the army are listed and it is described along what routes the grain could be transported from the granaries in the prefectures and districts to the war theatre. From this list it can be seen that from all the important grain-producing regions in Sichuan rice was shipped to Chengdu, Guanxian and Yazhou. From some places in the north grain was also directly shipped to Dzagu on the northern route. For the southern route, the grain was directly shipped to Yazhou, and not on to Chengdu. Along many routes it was possible to transport the grain by either waterway or overland (along the Minjiang, Tuojiang and Dadu rivers), and from a certain level of cost—or urgency—on it became more convenient or less expensive to transport it by road instead of by boat, especially when it came from the northeastern prefectures in Sichuan (with the districts of Yilong, Yingshan, Quxian and Dazhu). The rice originating from those prefectures therefore was not transported downstream along the Jialing and Qujiang rivers and then the Yangtse river up to Luzhou and Yibin from where it could be shipped

144 Figures according to Lai, *Qianlong zhongyao zhanzheng*, p. 305.

145 *Pingding Liang Jinchuan fanglüe*, juan 37, fol. 13b (QL 37/8/gengchen). Figures also in Lai, *Qianlong zhongyao zhanzheng*, p. 333.

146 *Pingding Liang Jinchuan fanglüe*, juan 67, fol. 6b–7a (QL 38/7/yisi).

147 *Pingding Liang Jinchuan fanglüe*, juan 13, fol. 26a (QL 36/12/gengyin).

148 *Jinchuan junxu li'an*, juan 2, fol. 123a–137a.

upriver again to Chengdu and Yazhou. Instead, the grain was taken overland to Chengdu directly, from where it was shipped on to the transport bases in Yazhou (and on to Dartsedo) or Guanxian. From the transport bases in Guanxian, Dzagunao and Dajianlu the rice was transported along the logistics routes to the camps (to be consumed by the troops) and to the logistics stations (to be consumed by the porters working there). The logistics stations were therefore also called *liangzhan* 'grain stations'.

There are three sources explaining where the grain came from: the first one is the listing of shipment routes from the prefectures of origin in the province of Sichuan, as described above. The second source are statements in official documents which give evidence that all rice originated in the province of Sichuan. The third source are data about the rice kept in the provincial granary in Chengdu, collected by Pierre-Étienne Will and R. Bin Wong.

Documents show that no rice was brought in from other provinces to feed the troops in Jinchuan. The province of Sichuan had produced enough rice over the last years so that there should have been absolutely no problem to obtain rice in the district granaries near Chengdu.[149] Because the harvest in the middle and lower Yangtse provinces had been excellent and those densely populated regions were not in need of grain supports from Sichuan. The emperor therefore strictly forbade exporting rice to other provinces, except for small-scale border trade from the prefecture of Kuizhou (modern Fengjie, Chongqing).[150] This measure would also help stabilizing the grain price in Sichuan. This export ban was only lifted after the fall of the main stronghold Le'uwei.[151]

Although it is known from the regulations and from documents that the Banner troops were better nourished than their colleagues from the Green Standard units because of their higher salted-vegetable pay, the quantities of rice were the same for both of them.[152] There also seemed to have been enough rice in the camps, and only when some units were cut off from their supplies by the enemy, or when at the beginning of the war the logistics routes were not yet fully established, soldiers, especially in winter, suffered from hunger. In one case, when rice was not sufficient, the responsible official only distributed half the ration due (0.401 *sheng* instead of 0.83 *sheng*), while the rest was paid out in 4 *fen* of silver, which

149 *Gaozong shilu*, juan 938, fol. 30b–32a (QL 38/7/yichou).
150 *Jinchuan dang* 38/III/00045 (QL 38/7/8), 00059 (QL 38/7/9).
151 *Jinchuan dang* 40/IV/00115–116 (QL 40/10/25).
152 *Jinchuan dang* 38/IV/00061 (QL 38/10/10).

at that time only sufficed to buy 'one single bun' (*bobo*). This procedure was an open violation of the common rules according to which it was not allowed forcing troops to accept silver instead of rice.[153] Never before had troops been obliged to accept money instead of their daily rations, a procedure only permissible when regular number of posts of manservants in the garrisons were left vacant, or for surplus military labourers. The responsible official accordingly received his sentence.[154] That the governmental regulations for expenditure and for rice rations were based on a reasonable calculation and did not just follow the simple objective of saving money becomes clear when looking at one case where it was revealed that soldiers tried to eat less rice and to sell what rice they had saved to merchants or to the logistics stations. This came to light when a rice merchant named Tian Jiguo was arrested who did not carry any rice with him but instead more than 2,700 *liang* and four delivery tickets for rice (*mipiao*), indicating how much rice he had to deliver at a fixed point. The officials assumed that this "evil-doing merchant" had tried purchasing rice somewhere from logistics stations and camps instead of at Chengdu, saving himself considerable transport cost. The emperor, furious not only about this merchant, but rather about this kind of craftiness among the troops, argued that the amount of one *sheng* of rice (about 1 litre) per person and day was a reasonable amount to nourish a labouring person and to give him enough energy for fighting the 'rebels'.[155] According to this objective calculation it should not be possible for the Green Standard and the provincial Banner troops to have any surplus rations. If troops obtained more than one *sheng* of rice per day and person, this was to be shared with the manservants, and not to be hoarded to make a profit. If the soldiers had any surplus from their daily rations, this could only mean that they were given more than necessary, which was the fault of their superiors in the logistics. Any surplus in the camps should be stored for later, in case of a shortage. Tian Jiguo was executed to demonstrate how serious his profiteering was considered, and the generals were ordered to keep track of all merchants in the region trying to do the same. For a moment it seemed even probable that a logistics official named Liu Zuceng managing the rice supply had tried to gain profit from such a business,

153 *Hubu junxu zeli*, juan 8, fol. 1b.

154 *Jinchuan dang* 39/III/00309 (QL 39/9/15), 00323 (QL 39/9/21).

155 The emperor asked why 0.83 *sheng* of rice was just sufficient to nourish a Bannerman, while the Green Standard soldiers were not able to consume this amount. *Pingding Liang Jinchuan fanglüe*, juan 57, fol. 25b–26b (QL 38/r3/dinghai).

before it was found out that charges against him were unjustified.[156] The misuse to sell surplus grain had already been recognized during the first Jinchuan war when overseers of the logistics stations sold their surplus to private hauliers to make a profit. The emperor also admitted that under the conditions found in Jinchuan it might well have been possible that there were several cases when grain was delivered to the wrong station, which automatically led to misappropriation.[157] In fact, there were also other cases of private persons being arrested for dealing with government-issued delivery tickets (*piao*) in the wrong places.[158]

Two other examples shall be brought forward to highlight the misuse of grain. Two grain logistics labourers named Zhang Wenyao and Wang Jiayu had illegally sold grain. In their station an amount of 8 *hu* of rice (actually not very much: it corresponds to a mere 414 litres) were missing, and their foreman called Yang Guoxing ordered them to buy rice to compensate for the missing amount. Yet when delivering the rice they claimed to have spent more than they actually had. This was found out, and all three deserted.[159] A much more serious case was that of the labourers Du Chaozhen, Liu Chaogui and Wang Deyu who sold 25 *hu* of grain (1,294 litres) by carving a faked official seal. The responsible grain manager (*liangyuan*), prefect (*zhizhou*) Zhang Keming, soon discovered this abuse of authority and was rewarded by a position in which he would immediately be promoted to a prefecture in the province of Zhili in case of a vacancy. The responsible station manager (*zhanyuan*), magistrate (*zhixian*) Yuan Wenmo was not punished because he had the delinquents arrested immediately.[160]

In their study of the state granary system Will and Wong provide some data on the provincial state granaries. From those data it can be seen that the quantities of rice stored in the main granary of the province of Sichuan over time can be represented by a parabolic curve which rose almost regularly from QL 20 (1755) on (1.7 million *dan*) to QL 50 (1785; 3.7 million *dan*). The figures from the first twenty years of the Qianlong reign vary considerably, and many data are missing. The great exception in this development is a sudden 'hole' in the stored rice quantities between the years QL 36 and 41 (1771–1776), which is exactly the period

156 *Jinchuan dang* 38/IV/00357 (QL 38/12/25).
157 *Pingding Liang Jinchuan fanglüe*, juan 59, fol. 13a–14a (QL 38/4/xinhai).
158 *Jinchuan dang* 40/III/00113 (QL 40/8/9).
159 *Jinchuan dang* 40/IV/00035 (QL 40/10/17).
160 *Jinchuan dang* 40/IV/00107–108 (QL 40/10/23).

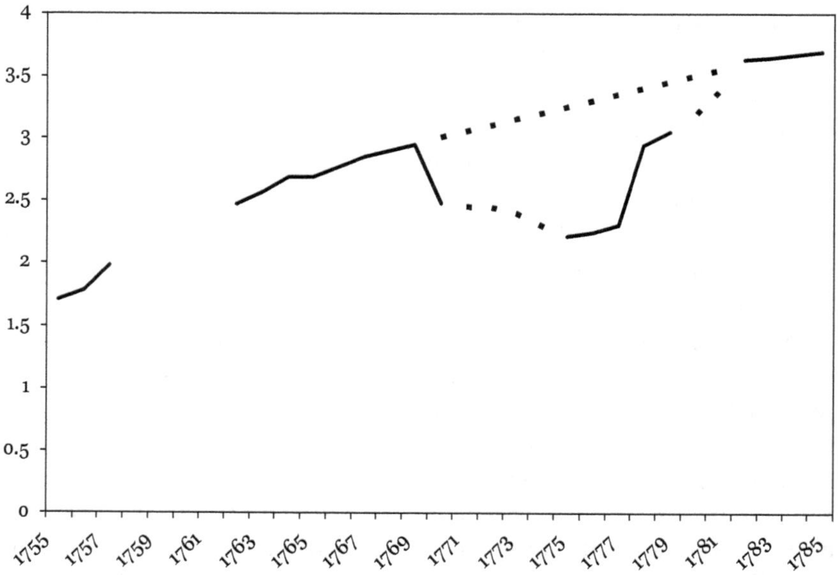

Source: Pierre Étienne Will and R. Bin Wong, Nourish the People: The State Civilian Granary System in China, 1650–1850 (Ann Arbor: University of Michigan Press, 1991), pp. 528–531. The solid line is existent data, dots are extrapolations. Time frame QL 20–QL 50.

Diagram 2. Civilian granary holdings (in million *dan*) of the provincial granary of Sichuan, 1735–1785.

of the second Jinchuan war. An extrapolation of the parabolic graph makes it possible to roughly calculate the amounts of rice 'missing' to fill the ideal graph (see Diagram 2, the area between the upper and the lower curve). This results in about 4.9 million *dan* of rice which in an ideal case could have been stored in the provincial granary, but instead went elsewhere. During the second Jinchuan war 2,963,527 *dan* (306,670 m³) of rice were consumed, which is even less than the 'missing' amount of rice. This shows that the state granaries in Sichuan were able to provide the total amount for the campaign and that no rice had to be imported from other provinces.

Apart from rice the troops and labourers consumed 31,902 *dan* (3,300 m³) of flour, and an amount of 79,302 *dan* (8,200 m³) of beans was consumed by the transport animals.[161] Flour was mainly used to make *tsampa* or cakes and rusk with, which was easy to transport and thus an ideal diet on the march, also in the West. Flour was otherwise used as rations when the troops had to spend a period of time without any access

[161] *Jinchuan junxu li'an*, juan 2, fol. 180a, 182a, 183a.

to supplies. For this purpose the flour was roasted (*huimian*, *chaomian*) for conservation.

The recruiting of labourers was undertaken according to the tax rate of the region liable for recruitment. For each 100 *liang* of taxes, between 2 and 10 labourers were to be supplied. As a compensation to the district, the *jintie* surcharge was suspended in districts which had provided labour force to the government. This way, about 150,000 labourers had been recruited from throughout the prefectures and districts of Sichuan.[162] The locations where the labourers came from were spread over the whole territory of the province of Sichuan, so that there was virtually no district where peasants were not to serve the army.

The 'barbarian affairs' files covering the logistics and financial matters of the first months of the war contained the following regulations concerning labourers:[163] From ten granaries in Sichuan,[164] 32,900 *dan* of rice had been moved, and 79 *dan* been purchased from private suppliers. When buying rice, the current monthly price officially reported to the local government had to be taken into account, which was about 1.1 to 1.2 *liang* per *dan*.[165] On the long road to the southern route and onwards to Djanggu on the Greater Jinchuan River, for each *dan* of rice two porters had to be used. There were also native porters recruited from among the local population by the native kings. The regulations for the logistics and the employees in the stations changed somewhat after the accounts were decided to be enclosed in the military expenditure files, from the end of QL 36 (1771) onwards. The main difference was that from then on, the rice was not supplied from state granaries but was to be bought on the market, which drove the cost up. According to the old files, for each *dan* of rice, only 1.35 *liang* was to be paid, including an allowance for transport losses (*jiahao*). The husked rice could be bought directly from granaries throughout the prefectures and districts of the province of Sichuan, but the new files postulated that the rice for military supply should be bought only in Chengdu. Therefore all rice rations first had to be shipped to Chengdu. The price in Chengdu was originally not supposed to be higher than 1 *liang* per *dan*, which is cheaper than the prices paid

---

162 *Jinchuan junxu li'an*, juan 2, fol. 71b–72a. Compare also the important article of Dai, "Qing State, Merchants", p. 49.

163 *Jinchuan junxu li'an*, juan 1, fol. 3b–5a.

164 Chengdu, Tongchuan, Ya'an (Yazhou), Jiading, Xuzhou, Zizhou, Mianzhou, Meizhou, Qiongzhou, Luzhou.

165 *Jinchuan junxu li'an*, juan 2, fol. 73a.

in QL 36 (1771) and also lower than the government conversion rate of 1 *dan* : 1.2 *liang*. In the years QL 39 (1774) and QL 40 (1775) the rice stored in the granaries was used up before the new harvest, so that the price for rice in Chengdu increased substantially and the grain had to be shipped from other prefectures directly to the transport 'terminals' in Guanxian and Yazhou, where it ranged between 2.3 und 3.3 *liang* per *dan*, depending on the place of origin.[166]

There is a very important statement in the *Jinchuan junxu li'an* which says that in the old files rice was exclusively moved by state-organized transport (*guanyun*), while according to the new files after QL 38/6/29 (Aug 17, 1773) everything was transported by private entrepreneurs (*shangyun*).[167] Yet already at a very early stage of the war, namely at the beginning of QL 37 (1772), the leading generals Wenfu, A'rtai and Agui suggested contracting private entrepreneurs for the transport of grain. This method had proved its usefulness during the first Jinchuan campaign, and it was apparently not much more costly to have entrepreneurs move the rice than to rely on coerced labourers, as the following document shows:

> [...] In the files of the war against Jinchuan during the years QL 12 and 13 [1747–1748], we see that apart from the state-organized transport, entrepreneurs have been recruited in order to support the state-organized transport in places where the latter was not sufficient, so that there were actually no reasons to worry about any deficiencies in supply. Since Your servant [Wenfu] has been in charge, he has diligently researched and found out that the grain price is low and people therefore jump about for joy. Since several months until now (QL 37/2, Mar 1772) we have made use of labourers recruited from among the people to take care of the transport. Yet Sichuan is a very large province, and of the various prefectures and districts, some are located near, but others far away [from the war theatre]. And for the soldiers fighting on the three routes, tens of thousands of porters are used, especially since several thousand new troops have been dispatched and arrived from the provinces of Guizhou and Shaanxi, so that in each transport station the porters have to carry additional amounts of loads. Yet it is not possible to dispatch so many new porters at the same time from throughout the province [of Sichuan], and to complicate matters further, springtime has just begun, when the peasants have to work their fields [and are therefore indispensable in their home villages]. [The labour in the logistics stations can thus] only be fulfilled with great difficulties. Moreover, the logistics stations on all three routes, numbering between 10 and 13 [on each route], have to be supplied with daily provisions for the porters to be delivered by each

[166] *Jinchuan junxu li'an*, juan 2, fol. 73–74.
[167] *Jinchuan junxu li'an*, juan 2, fol. 74b.

> station. The consequence is that from each *dan* delivered to the encampments not a small amount has to be deducted [to feed the porters].
>
> Concerning the transport by private entrepreneurs, the transport cost might [in pure figures] be more expensive than the transport organized by the state. Yet for the entrepreneurial transport we do not have to procure any provisions, and compared with state-organized transport in what has to be paid daily [in rice or provision pay], the overall expenditure is also not that much [i.e. not more than for entrepreneurial transport].
>
> If we add privately organized transport to the [state-drafted] porters already transporting [the grain] from station to station, this would greatly benefit the people because the peasants would not be obstructed in their work; and it would verily be of great advantage to the provisions for the troops in the encampments. Both sides, the people and the army, would therefore benefit [from such a method…].[168]

One of the greatest problems of state-organized transport was that the government could not endlessly exploit the labour-force of peasants—it had, according to the regulations, to replace (*huanban*) the whole contingent every three months. Thus at a certain point of time, thousands of people had to be exchanged in situ, with the consequence that while the old contingents were still working, new contingents had to march to Jinchuan and to take over the work immediately in order not to cause any delays in the grain transport system. There is one example in the memorials in which Guilin complains that of the 11,800 new workers, only some 500 had have made their way to the southern route. Therefore, Wenshou in QL 39/2 (Mar 1774) suggested prolonging the duration of service from 3 months to 5 months.[169] If this system was correctly adopted, 11 replacements of personnel on service for three months should have taken place, and there should have been four periods during which the personnel served five months. The service periods were adhered to strictly so that even at the end of the campaign people working in a logistics station being broken up had to wait in one of the nearest stations until the end of their service period before being allowed to going home.[170] On the other hand there was a regulation in the old precedents which said that the labourers were not to be given any additional family allowance when they had to stay longer than three (resp. five) months. After the new precedents became applicable, many porters may have served longer than their actual service

168 *Pingding Liang Jinchuan fanglüe*, juan 21, fol. 14b–15b (QL 37/2/renchen). Similar arguments can be found in another memorial, recorded in juan 24, fol. 9b–9b (QL 37/3/jiazi).

169 *Pingding Liang Jinchuan fanglüe*, juan 92, fol. 8b–9b (QL 39/3/xinwei).

170 *Jinchuan junxu li'an*, juan 2, fol. 70b–71a.

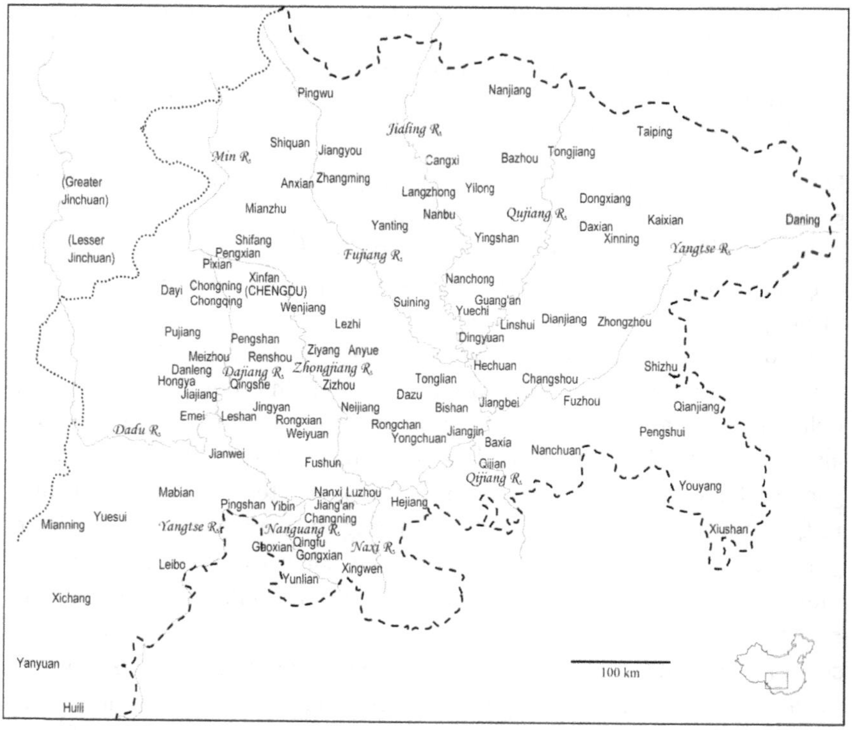

names of districts

Source: *Jinchuan junxu li'an*, juan 2, fol. 71b–72a. Map created by the author, based on Tan, *Zhongguo lishi ditu ji*, pp. 39–40.

Map 4. Districts in Sichuan from which peasants were recruited for military labour.

period would have required them to do.[171] It is not known if each of the four rounds, during which peasants from Sichuan were recruited according to the tax registers, constituted one such period.[172] There were 462,097 people working as porters during the Jinchuan campaign, about 151,800 of whom were recruited according to the tax registers at the beginning of the campaign.[173] In each station, there were between 500 and more than 1,000 porters, or 750 on average, which means that towards the end of the campaign, when a very high number of logistics stations was in operation, about 250,000 porters were permanently present in Jinchuan, more than twice the number of soldiers serving at the front.

The biggest advantage of private transport was that the organizers of the logistics did not have to supply rations for the porters nor any other food like vegetables and meat, for the preparation of which the logistics officers had field kitchens installed where the soldiers and porters could buy food and other commodities of all kinds. People hired by the entrepreneurs undertaking the transport did not have to be paid any monthly wages, nor any family allowances. Yet the maintenance of the 353 logistics stations was still in the hands of the government.

The factor by which the transport price per *dan* increased when switching over from state-organized transport to 'outsourced' transport is about 4, which looks as if commissioning a private cargo entrepreneur was four times as expensive as coercing peasants to work for a pittance. At first sight this seems plausible, but, why then, did the leading generals vote for private transport with the argument that cost was almost the same? Firstly, the rice is included in the cost for the outsourced transport, which could be more than 2 *liang* per *dan*, but less, when the state bought this rice with its own prescribed prices, at 1.2 or 1.35 *liang* per *dan*. Secondly, family allowances (*anjia yin*) did not have to be paid for people working for an entrepreneur. Thirdly, while the state had its own standards when paying labour and allotting daily rations, it had to pay enough to make sure the peasants would not run away, which means that the original prices as paid during QL 36 (1771) cannot be taken as standard pay for the whole campaign. One example might elucidate this matter: When 0.05 *liang* were the daily pay (or transport cost) inside provincial territory,

---

171 *Jinchuan junxu li'an*, juan 2, fol. 68a.

172 *Jinchuan junxu li'an*, juan 2, fol. 71b–72a. Dai, "Qing State, Merchants", p. 49.

173 *Jinchuan junxu li'an, Zonglüe*, fol. 3b; juan 2, fol. 185b. Dai, *Sichuan Frontier and Tibet*, p. 181, interprets the figure of 462,000 as the number of 'contracts' and not as that of individuals, and so obtains a figure of 360,000. See also Dai, "Military Finance", p. 308.

and 0.08 *liang* outside,[174] the rice price in Dzagunao was already 2.692 *liang* per *dan* by official transport from Chengdu, and the price in Somo 5.268 *liang*, when the rice price in Chengdu has been 1.35 *liang*.[175] Travelling to Somo, for example, took about one month, when carrying rice. This would mean that the daily expenses for the transport alone would be 0.13 *liang* for state-organized transport (with a rice price per *dan* of 1.35 *liang* at the point of destination), and daily expenses of 0.31 *liang* for the private transport (with a rice price of 2.67 *liang*). Yet the crucial point is that the state would not be willing to pay a price three or four times as high as the cost for the 'normal' method of using labour conscripts. The price for state-organized transport must have become much higher during QL 37 (1772) and QL 38 (1773), which becomes evident with the following regulation: The conscripts did not only receive a family allowance paid in one lump sum, but instead were given this amount monthly, in order to make them stay with their work.[176] Instead, as a once-and-for-all gratification of 2 *liang*, they were given 3 *liang* monthly, which is a fairly good pay. As this amount was calculated on a day-by-day basis, the labourers were entitled to an extra payment for every day on top of their stipulated service period. This means that two labour conscripts, for example, peasants from a prefecture far away from Jinchuan, were probably paid 30 *liang* together, 12 of which were just paid for their way to Chengdu and back. During their period of service to Somo, they were able to carry two and a half *dan* of rice (when walking all the way back). This means that the transport of one *dan* of rice cost 12 *liang* (in the worst case; and excluding cost for the rice itself), and even for people living near the border, the transport must have cost a lot more than 3.2 *liang* per *dan*, which was the price effective at the beginning of the campaign. They represent the direct cost of state-organized transport before the commanders totally switched over to 'outsourced' transport of rice. But surely the organizational consequences and their indirect cost also played an important role: recruiting peasants, guiding them to Jinchuan, supervising them, feeding them, paying them and finding new staff when dozens of peasants deserted—all these activities likewise consumed time and labour force. The consequence of these

[174] *Hubu junxu zeli*, juan 5, fol. 1a–1b.

[175] This calculation does not include the family allowance for the two porters for the one *dan* of rice.

[176] *Jinchuan junxu li'an*, juan 2, fol. 71a–b. Dai, "Qing State, Merchants", pp. 52, 55.

considerations is that the 933,227 *dan* of rice shipped by state-organized transport, were much more expensive than the regulations made out.[177]

This is an important hint indicating that the sums officially required according to the regulations for war expense like the *Jinchuan junxu li'an* when settling accounts were often so far below those actually paid by the war logistics bureau that officials over and over again complained of high prices for transport and goods to be bought on the open market. This went so far that they finally proved to the Ministry that it was no more costly, but even more efficient as far as the organization was concerned, not to transport rice by state agencies but by haulage contractors. A market price of about 2 *liang* per *dan* was already too high for the purchase of large amounts of rice, as the Ministry criticized.[178] The daily consumption of rice by porters (not including the consumption by troops and civilian managing staff) along the three main logistics routes was more than 700 *dan* (72 m^3). For this reason it was suggested to use more roast flour and to hand out roast flour every fifth day instead of rice, in order to somewhat cut down the consumption of the latter.[179] Facing the alarming rice prices of 2.4 to 3.6 *liang* in Chengdu, 5 to 6 *liang* in Long'an, Maozhou and Yazhou and up to 10 *liang* in Dartsedo, the logistics managers had indeed every reason to deliberate about a higher amount of the cheaper wheat and barley to be given to the troops instead of the expensive rice.[180] The government tried balancing the official price of rice in a range between 1.35 *liang* and 1.85 *liang*, which is substantially lower than the real market prices. The difference can be explained with the fact that the price indicated is that of the place of origin (a net price), for example, in Chongqing in the far east of the province of Sichuan, not taking into account that the rice had to be transported to Chengdu and Yazhou, which led to a sharp increase in the gross price. The rice shipped to Chengdu, for example, cost

---

177 This amount is the difference of 2,030,300 *dan* of rice transported by private contractors (itself being a sum of the figures in *Jinchuan junxu li'an*, juan 2, fol. 77a–79b) and the total sum of 2,963,527 *dan* of rice.

178 *Pingding Liang Jinchuan fanglüe*, juan 70, fol. 12b–13b (QL 38/7/xinyou). Li Zhongqing, *Zhongguo xinan bianjiang de shehui jingji: 1250–1850* (Beijing: Renmin chubanshe, 2012), 238–239, provides lists with market prices of rice in border regions during the respective time, which were about 2 *liang* per *dan* in Yunnan and 1.2 *liang* per *dan* in Guizhou. Price fluctuations from the province of Zhili can be found in Lillian M. Li, "Grain Prices in Zhili Province, 1736–1911: A Preliminary Study", in *Chinese History in Economic Perspective*, ed. by Thomas G. Rawski and Lillian M. Li (Berkeley: University of California Press, 1992), pp. 77–79, 91–92, such of Gansu in Perdue, "Qing State and Gansu", p. 114.

179 *Jinchuan dang* 40/I/00113 (QL 40/2/8).

180 *Jinchuan dang* 40/III/00139 (QL 40/8/12).

2.31 *liang* per *dan* when transported from Jiading (Leshan) and 3.374 *liang* per *dan* when originating in Chongqing in the far east,[181] the transport cost amounting to 24 per cent (resp. 48 per cent) of the gross price. While the state granaries stored rice as well as flour, the beans were exclusively bought on the private market.

Flour was likewise transported not only by government-paid porters but also commissioned to private hauliers. The price offered by private hauliers likewise turned out to be cheaper than the cost for buying and transporting the flour from Sichuan to the camps by government-hired porters.[182]

Of the 2,963,527 *dan* of rice transported to the logistics stations and the camps, only a very small amount did not origin in the 'ever-normal' state granaries (*changping cang*) or the private market. These was grain from the military post (*xun*) of Changtang that had contributed 19,069 *dan* (1,974 m³) of rice, as well as 24,877 *dan* (2,575 m³) of rice from semi-private 'charity granaries' (*yigu* 'charity rice'), and rice from the garrison in Xilu (i.e. Dartsedo), which were not included in the regular accounts.[183]

There were at least two incidents when grain burnt: In the camp of Dangga over 2,000 *dan* were lost, for which general Agui was responsible. Yet because his men had captured more than 3,000 *dan* of grain and large amounts of gunpowder, muskets and cannons from the enemy, the emperor pardoned Agui.[184] The grain station manager Zhang Shilu, whose carelessness had led to the burning of 400 to 500 *dan* of rice in a station in Okshi, was not spared and lost his office.[185] At the end of the war 121,862 *dan* (12,613 m³) of rice were left over and had to be stored in the military colonies (*tunfang*) established in the Jinchuan area.

The Transport of Money

Money was mostly paid out in the shape of chunks from silver ingots, due to the simple fact that silver was much easier to transport than copper cash. The conversion from silver crumbs to cash was then undertaken by private money-changers in the camps who calculated with an actual market price. That silver was taken along with the troops can also be

181 *Jinchuan junxu li'an*, juan 2, fol. 73–74.
182 *Pingding Liang Jinchuan fanglüe*, juan 79, fol. 15a–16a (QL 38/10/jiwei).
183 *Jinchuan junxu li'an*, juan 2, fol. 181b.
184 *Jinchuan dang* 38/III/00448 (QL 38/9/1), 00453 (QL 38/9/1).
185 *Jinchuan dang* 38/IV/00289 (QL 38/12/12), 00329 (QL 38/12/20), 00341 (QL 38/12/24).

seen from the large amount of silver (more than 56,000 *liang*) which was lost during the battle of Mugom: It was also much cheaper to transport the silver sheaths to the camps instead of large numbers of copper cash strings (*guan*).[186] At the beginning of the second Jinchuan war, pressed for time, it had been custom to pay out the daily provisions inside the country in cash, namely 30 *wen* (coins, per day) for 0.0135 *liang* of silver, in total accruing to a sum of 2,624 *liang* in that case.[187] The reason for this procedure had been that the dispatch of the troops was so urgent that they were in fact equipped with 3,000 cash each (corresponding to the provision of 100 days) to take with them. The troops would then be able to buy their food from the sutlers without undergoing a time-consuming transaction with the money-changers. When drawing up the respective account, of course, silver had to be used as calculation currency. But the Ministry, although allowing such a procedure to the war logistics bureau, did not want to pay any sums surpassing the regular amount of provision pay. A part of these sums therefore had not been brought to account at all (probably because of conversion problems) and was later waived by the emperor. It could be seen in the earlier parts of this chapter that also during other campaigns troops were handed over cash, especially during their march to the battlefield.

The procedure at the beginning of the war, in the summer of 1771, was as follows: the soldiers, while on their way from the garrison to the front, were daily given 0.0135 *liang* of silver which bought one *sheng* of rice. The money was disbursed in the garrison for the number of days the soldiers would need to leave provincial territory, which means that they took along the money to buy their food on the way. Upon leaving the district of Guanxian or the prefecture of Yazhou, entering territory 'outside the country' they were given a daily ration of 1 *sheng* of rice (1.035 l), or 1 *jin* (597 g) of flour. The officers were not given any money to buy food with, as long as they were within the imperial territory. Yet once past the border of the districts and on entering the territory of the native kings, they were allotted 1 *sheng* of rice per day, too, irrespective of rank. This practice was based on the idea that while a common soldier had to be taken care of by

186 56,000 silver *liang* have a weight of 3,500 *jin*, or 2,090 kilos, while 56 million cash coins have a weight of 224,000 kilos, with an average weight of 4 grams per coin, according to David Hartill, *Qing Cash* (London: Royal Numismatic Society, 2003), p. 28, or Vogel, "Chinese Central Monetary Policy", p. 11.

187 *Jinchuan junxu li'an*, juan 1, fol. 3a; juan 2, fol. 147a. This corresponds to a market exchange rate of 2,222 *cash* : 1 *liang* instead of the official rate of 1,000 : 1, showing how much cash could depreciate during the late Qianlong reign.

the government, an officer as a state official had an income, which was supposed to be high enough to feed him, as long as he was on imperial territory.

Considerable amounts of money had to be transported to the camps and the logistics stations, silver as well as copper cash.[188] The money sent to Sichuan either from the treasury of the Ministry of Revenue, from the various provincial treasuries, or from the hands of entrepreneurs, was not going via a bank account or a similar facility making it possible to instantly transfer money to a distant place where the sum is immediately made available by a corresponding bank. Instead, the silver was physically transported from Beijing or the various provinces to Sichuan. There were special cases or 'sheaths' (*qiao*) in which the silver was packed and which were transported by a heavily protected convoy. According to the regulations, such a sheath package contained one thousand *liang* of silver bullion, with an actual weight of 62.5 *jin* (37.3 kg).

A precious item like silver bullion could only be placed into the hands of trustworthy people, and therefore each province sending money had carefully to select their own reliable officers who ensured a safe transport. On QL 39/2/30 (Apr 10, 1774) the acting Left Vice Minister of the Ministry of Revenue, Gao Pu, asked for a better protection of the silver transports and proposed that sums over 100,000 *liang* were to be guarded at least by an assistant brigade commander (*shoubei*) or, better, by a brigade commander (*youji*), each of them at the head of several hundred soldiers.[189] To look after such a transport was an everyday task of the military garrisons in the provinces.[190] The dimensions of such an enterprise were indeed considerable: The transport of 100,000 *liang* of silver required 50 mules or horses that were led, as Gao Pu complained, by too many people, a large part of whom were labour conscripts who did not really care for that job. Chen Huizu, acting governor-general of Hu-Guang (Hubei and Hunan), on the other hand had great trust in his military personnel and advocated that also a company commander (*qianzong*) or even a squad leader (*bazong*) could be entrusted with a transport of more than 100,000 *liang* silver.[191]

It had not always been the rule that the province of origin alone was responsible for the transport. In the case of money sent from the two

---

188 *Jinchuan junxu li'an*, juan 2, fol. 65b.

189 *Gaozong shilu*, juan 953, fol. 20b–21a (QL 39/2/guichou).

190 Luo, *Luyingbing zhi*, 252.

191 *Gaozong shilu*, juan 981, fol. 25a–25b (QL 40/4/bingwu).

provinces of Hu-Guang on QL 37/6/24 (Jul 24, 1772), personnel from Sichuan was to be dispatched to Wuchang or Changsha to take charge of the money and to transport it to Chengdu. The selection of trustworthy transport officials was probably also undertaken by the Sichuan officials. In all other cases, personnel from the place of origin—as far as we know—cared for the transport.

There is at least one case when money went missing during the whole long war.[192] On QL 39/3/18 (Apr 28, 1774) it was reported that one sheath of silver was missing in a smaller transport on the new western route.

The regulations for military expenditure, the *Junxu zeli*, which were issued in QL 49 (1784), contain a specific chapter dealing with the transport of silver bullion from the treasury of the Ministry to the point of destination.[193] The transport conditions of silver sheaths containing pay for the army were basically the same as for other objects, with a daily distance of 100 *li* in flat territory, transported preferably by carts, otherwise by animals, or porters in the worst case, the daily minimum distance of which was 40 *li*. According to the Jinchuan precedents, porters carrying pay for the army were not only paid a regular porterage, but also—instead of rice—pay for daily provisions when walking back to their home station without load (*huikong*). This was an extraordinary gratification, which was forbidden to be continued as a permanent precedent. When money was disbursed from the treasury of the Ministry of Revenue in Beijing, the province of Zhili had to send an especially dispatched grand official from a circuit or a prefecture (*dao-fu dayuan*) who, together with a secretary (*siyuan*) dispatched by the Ministry, checked the weighing and packing of the silver sheaths. When they reached the district of Daxing or Wanping in the outskirts of Beijing, carts had to be kept ready to be loaded with the money to be transported. An experienced official from one of the sub-prefectures around Beijing (*tingyuan*) was to be dispatched who held himself ready at the first logistics station to accomplish the first handing over from the metropolitan prefecture to the province of Zhili. For each subsequent station, the governor-general selected and dispatched one member of the circuit administration (*daoyuan*) and one Banner regimental vice commander (*fu canling*), who both supervised the transport, and he also selected and dispatched one brigade commander (*youji*) and vice brigade commander (*dusi*) from the prefectures or sub-prefectures on the way to

192 *Gaozong shilu*, juan 956, fol. 18a–18b (QL 39/4/guisi).
193 *Hubu junxu zeli*, juan 5, fol. 4a–5b.

protect the transport with their troops from station to station. This way, the silver was relayed on until it reached the treasury (*fanku*) in the province of destination. For each handing over, the time had to be reported to the Ministry of Revenue, when the silver had entered or left the respective territory. Before 1774, all silver being disbursed from the Ministry of Revenue to be transported to another province, had been accompanied by a secretary from the Ministry (*siguan bithesi*), who took over responsibility on the long way to the destination. When the money was finally handed over to the province of destination, it was repeatedly found out that on the way some sheaths had gone missing. For this reason, the Ministry of Revenue suggested a tighter control that was put into the hands of the governor-general of Zhili, who was the first provincial official to take over the money.

The regulations in the *Jinchuan junxu li'an* provide some additional information, as the *Junxu zeli* regulations almost exclusively deal with the situation in the Capital and not with that on the way.[194] When transporting silver ingots, each 1,000 *liang* were to be stowed in a sheath, and, as the regulations say, by no means in a basket or other containers. Nevertheless, statements in the same paragraph can be found that openly contradict this regulation and clearly talk of the use of baskets for the transport of contribution silver (*juanyin*). The prescribed sheath was transported by two porters from the regular courier stations of the districts and prefectures. Silver was much too precious to leave it in the hands of porters hired on the free labour market, as it is said, and only in urgent cases, when there were not porters enough in the transport or courier stations, porters could be hired and were to be paid according to the regulations for hired workers. Especially in cases like these the responsible official had to keep an exact account of the number of labourers, the distance they covered, and the amount they were paid. For a distance of 100 *li*, which is a normal one-day march, on their way to Chengdu, each person was to be given 0.1 *liang* of pay. In regions, where no regular transport or courier stations existed, the porterage had to be given according to time instead of distance, and was to amount to 5 *fen* per day, which means that in such a case the normal distance covered in one day was about 50 *li*.

When merchants or other people delivered contributions during contribution campaigns, the money had to be transported according to the above regulations from the province of origin to that of Sichuan. The

194 *Jinchuan junxu li'an*, juan 2, fol. 65a–65b.

money was then drawn from the account of the province of origin and transferred via the account of the Ministry of Revenue to the account of the province of Sichuan from where it was immediately paid out to the war chest. For this procedure the province of Sichuan had to draw up an account. Each 1,000 *liang* of contribution money had to be transported in baskets or wrappings of cloth, which were tied up with ropes to form a sheath containing the silver ingots. Instead of porters, pack animals could be used, each of which transported 2,000 *liang* of silver. The cost of transporting silver were, for example, from Chengdu to Ludingqiao 5 *liang* per load, to Taoguan and Dzagunao 3 *liang*, which was actually a little bit more than allowed by the precedences. Alternatively, but especially in precipitous territory outside the provinces, where these animals could not be used, native transport animals (*ulaa*) could be hired or alternatively two native porters instead of one horse. A longer commentary on this passage makes it clear that outside provincial territory, 'one thousand' *liang* instead of 'two thousand' had to be read in some documents, with the consequence that animals had to carry only 1,000 *liang* of silver and persons only 500 *liang*. This resulted in much higher cost for porterage than actually allowed for, because for each 1,000 *liang*, two persons had to be paid instead of only one porter. Yet under such conditions, each person would still have to carry a load of about 18 kilos.

Gunpowder

Gunpowder was produced and stored in the garrisons from where it was shipped to the war theatre as required. The province of Sichuan itself did not have large reserves of nitre (*huoxiao*). Due to the prolonged war in Jinchuan the gunpowder of the garrisons was soon exhausted and therefore either more had to be produced to replenish the garrison stores or to be bought from other provinces. The governor-general of Shaanxi and Gansu therefore dispatched a lower official to Hunan to purchase 200,000 *jin* (119 metric tons) of gunpowder to be taken to Xi'an to replenish the depots, and eventually to be transported on to Jinchuan.[195] In July 1772 A'rtai had repeatedly asked for additional gunpowder to be sent to Sichuan from other provinces.[196] All of the 200,000 *jin* of gunpowder kept ready in the Sichuan garrisons had been consumed, and therefore in September 1771, and again, in December 1772, he had requested to support the

---

195 *Pingding Liang Jinchuan fanglüe*, juan 67, fol. 3a (QL 38/7/bingwu).
196 *Pingding Liang Jinchuan fanglüe*, juan 11, fol. 19a (QL 36/11/bingyin).

war with 60 per cent of the gunpowder stored in the provinces of Hubei, Hunan, Shaanxi and Guizhou. Shaanxi sent 95,500 (57 t), Guizhou 101,900 (61 t), and Hubei and Hunan 20,000 *jin* (12 t) of gunpowder as well as bullets for muskets. Since the nitre produced in the state-owned workshops in Sichuan was not sufficient, Guizhou had helped out with 10,000 *jin* (6 t) of nitre. But not even this was enough, and A'rtai therefore asked for some more 10,000 *jin* of nitre to be sent from Yunnan, Shaanxi and Gansu. Le'rjin, governor-general of Shaan-Gan consented to dispatch 10,000 *jin* of nitre from Shaanxi, as well as gunpowder from the garrison of Shaanxi containing the same quantity of nitre. The Shaanxi gunpowder was to be used to substitute gunpowder from Gansu, a province which at that time was not able to provide such large amounts of gunpowder. Things therefore became much more complicated than should be expected: not only did Sichuan have to pay back money to other provinces, but in some cases even a third party was involved which later would have to be compensated. In July 1772 A'rtai ordered a further support by 80,000 *jin* (48 t) of gunpowder from Hu-Guang and 10,000 *jin* (6 t) of nitre from Guizhou. The transport cost for those items had to be brought to account in the provinces concerned.[197] In case of shortage it could even be that the logistics managers suggested using deteriorated nitre (*zouxiao*) in order to save cost.[198] If such methods were common, there is no wonder why the guns were so ineffective. The final account in the *Jinchuan junxu li'an* at least gives exact data about provincial transfers of gunpowder, which consequently had to be paid back by the province of Sichuan to the provinces of origin: Shaanxi provided 564,600 *jin* (337 t) of gunpowder, Shanxi 150,000 *jin* (90 t), Guizhou 497,000 *jin* (297 t), the provinces of Hubei and Hunan 941,000 *jin* (562 t), and Yunnan 126,000 *jin* (75 t). Hubei furthermore provided 200,000 *jin* (119 t) of nitre, and Guizhou 620,000 *jin* (370 t).[199]

There were several places in the Jinchuan territory where sulphur (*liuhuang*) and nitre could be produced, although only in small quantities. They were to be guarded in order to prevent the rebels from exploiting those sources.[200] The Lesser Jinchuan rebels also bought bullets and gunpowder from the other native kings, which could not be prevented,

197 *Pingding Liang Jinchuan fanglüe*, juan 32, fol. 20b–21a (QL 37/6/yiyou).

198 *Jinchuan dang* 39/IV/00059 (QL 39/11/1).

199 *Jinchuan junxu li'an*, juan 2, fol. 184a.

200 *Pingding Liang Jinchuan fanglüe*, juan 32, fol. 21a–22a (QL 37/6/yiyou); juan 36, fol. 18b–19a (QL 37/8/jisi); juan 60, fol. 18b–19a (QL 38/5/bingzi); juan 62, fol. 14b (QL 38/6/bingwu).

although the private trade of sulphur and nitre by Chinese and natives was strictly forbidden.[201] At least there was the expectation among the generals that after the final conquest of the two Jinchuans, sulphur and nitre sources could be exploited and used for the production of gunpowder for the post-war occupation troops. This would, when the gunpowder would be produced locally, save a lot of transport cost.[202] The leaders of the Jinchuan rebels even taxed their own population for gunpowder and collected 1 *jin* (597 g) of nitre and 2 *liang* (here used as a weight unit: 75 g) of sulphur from each household. This shows that every single fortified household in Jinchuan was well-equipped with weapons and gunpowder and that those materials were readily available.[203]

Metals

For the first few occasions of inimical encounters the troops took with them a certain amount of bullets and gunpowder. Lai Fushun quotes from a document saying that each soldier carried with him 1 *jin* (597 g) of gunpowder and 40 bullets. The *Jinchuan junxu li'an* includes a list of labour cost for the production of bullets of different sizes, ranging from 2.8 *qian* (10.5 g) to 16 *qian* (59.8 g).[204] The amount of lead for bullets was so immense that the troops soon ran out of raw material for casting their ammunition. Some 10,000 *jin* (6 t) of lead produced in the mines or workshops near Yazhou were sufficient to support the army for a while.[205] Yet in November 1773 Wenshou asked for 400,000 *jin* (239 t) of lead to be sent from the mines in Zhazi in the prefecture of Weining in Guizhou. The lead from this mine had originally been earmarked for the mints in Hunan and Beijing.[206] Therefore the war chest in Sichuan must have paid back the corresponding sum of money to the provincial mint in Hunan and the mints in Beijing later. The total consumption of lead for bullets was 2,525,086 *jin* (1,507 t).[207]

201 *Jinchuan dang* 40/I/00293 (QL 40/3/27).
202 *Pingding Liang Jinchuan fanglüe*, juan 48, fol. 15a–15b (QL 38/1/gengxu).
203 *Pingding Liang Jinchuan fanglüe*, juan 112, fol. 13a–13b (QL 40/1/gengwu).
204 *Jinchuan junxu li'an*, juan 2, fol. 45a–45b. *Hubu junxu zeli*, juan 1, fol. 2b–3b. Additional information on the production of bullets can also be found in (Qianlong 58) *Gongbu zeli*, juan 26, and (Jiaqing 20) *Gongbu zeli*, as *Gongbu zeli san zhong* in *Gugong zhenben congkan*, vol. 294–298 (Haikou: Hainan chubanshe, 2000), juan 32.
205 *Pingding Liang Jinchuan fanglüe*, juan 32, fol. 20b–2a (QL 37/6/yiyou).
206 *Pingding Liang Jinchuan fanglüe*, juan 79, fol. 3b–4a (QL 38/10/renzi).
207 *Jinchuan junxu li'an*, juan 2, fol. 184b. The text erroneously writes *qian paozi* 'cannonballs of lead'. The original units are: 2,525,086 *jin* 14 *liang* 2 *qian* 4 *fen*, with a hexadecimal

Without artillery it would have been impossible to destroy the infamous war-towers (*diao*) of the castles which were scattered all over the landscape of Greater and Lesser Jinchuan. The war-towers were so effective that the batteries often had to barrage night and day before the fortifications were ready for an assault.[208] Yet to transport the heavy pieces to the batteries (*paotai*), or constructing the batteries proved far more difficult than anticipated. Even aiming the pieces at their targets was far from being straightforward, as can be seen in the special dispatch of a Jesuit geodesist, Felix da Rocha, who was sent to support the battery staff.

Light cannon types with a barrel weight of more or less 100 *jin* (59 kg) could be transported on the back of a mule or by two porters, but the only possibility to move heavier artillery was to melt them down and to recast them near the front. Wheeled cannon carriages that allowed greater mobility in the field were known in China thanks to the Jesuits but were rarely used. The copper engravings and battle paintings of the eighteenth century mostly show cannon mounted on carriages with very small wheels, or no carriages at all. The casting of cannons on the spot in a field foundry (*suiying paoju*) was not only seen in the two Jinchuan campaigns. During the wars in the Western Territories cannons were cast in the western parts of the province of Gansu, from where they were transported, on camel backs, to the war theatre. During the Myanmar campaigns, too, cannons had been cast near the war theatre.

All materials for cannons and equipment, from gun metal to cleaning shovels,[209] had to be procured in Chengdu and to be transported by porters or mules to the camps. Iron for cannonballs and also for barrels was bought on the free market. Brass was procured in Chengdu from the offices that were in possession of brass or other copper alloys, namely the provincial mint (*qianju*) where the copper coins for the province of Sichuan were cast.[210] From the copper mines in Yunnan the whole country was provided with metal for casting coins, and therefore it went almost without saying that part of the metal transports could be deviated from Yunnan directly into the war region, without the detour through Chengdu.

---

jump from the unit *jin* 'pound' to the unit *liang* 'ounce'. The last three digits correspond to 0.89 *jin*. Compare Appendix 3.

208 *Pingding Liang Jinchuan fanglüe*, juan 11, fol. 18a (QL 36/11/bingyin); juan 12, fol. 6b (QL 36/12/gengwu).

209 A list of such equipment can be found in *Jinchuan junxu li'an*, juan 2, fol. 58b–59a, 63a–64b.

210 *Pingding Liang Jinchuan fanglüe*, juan 48, fol. 18a–19a (QL 38/1/xinhai). Chinese coins were not minted but cast.

The practice to confiscate the brass destined for the provincial mint was so common that the *Jinchuan junxu li'an* contains a special paragraph dealing with these proceedings: "Concerning brass, the metal is directly assigned from the mines in Yunnan, and for each amount [transported to Jinchuan] a particular memorandum has to be made."[211] The officers in charge thus borrowed the cannon brass from the coinage office and had to render this material back later when the campaign was over. A similar procedure took place for the lead that was used to cast the ammunition for muskets. The iron used for the cannonballs was not worth being reclaimed after battle. The staff of the camp foundry was, as the *Jinchuan junxu li'an* postulates, to be recruited from the logistics stations. These persons did not only bring the material for the cannons—brass bars or rods or 'crab shells' (*xieke*) produced in Leshan[212] from Chengdu to Jinchuan but also had to transport the finished barrel along with the ammunition from the camp foundries to the battery and thus provided the gun crew with new supplies. Head of the camp foundry was a foreman or 'engineer' who had to supervise the cannon casting, which means that he was a skilled workman and an expert in metallurgy. The greatest distance between the foundry and a battery was, as can be learned from the descriptions of the *Jinchuan junxu li'an*, up to ten stations, or day-travels.[213]

Conclusions

In his recent book on the southern inspection tours of the Qianlong emperor Michael G. Chang has at great length compared the travels of the Qing rulers to the far away cities of the Yangtse delta with military enterprises and imperial hunts, especially from the aspect of displaying martial prowess, but also concerning the field of logistics. Although Qianlong, unlike his grandfather, the Kangxi emperor, never personally conducted a campaign, he spent one-third of the year on the road during each of his six famous tours to southeastern China, accompanied by a "small army" of 3,000 people or even more.[214] The "prosaic problems" of transforming the Qing court into a mobile entity required techniques differing in nothing

211 *Jinchuan junxu li'an*, juan 2, fol. 46a–46b.
212 *Pingding Liang Jinchuan fanglüe*, juan 58, fol. 7a–7b (QL 38/4/xinmao).
213 *Jinchuan junxu li'an*, juan 2, fol. 44a–45b.
214 Michael G. Chang, *A Court on Horseback: Imperial Touring and the Construction of Qing Rule, 1680–1785* (Cambridge, Mass.: Harvard University Press, 2007), pp. 114–115.

from those for a military campaign. One inspection tour involved a total of forty-seven daytime rests, fifty-six main overnight encampments, and fourteen temporary places.[215] All these figures are comparable to those of the early short Western campaigns against the Dzungar leader Galdan. The State Council (*junjichu*) was responsible for coordinating the preparations of the enterprise, luckily an institution whose ministers had ample experience in organizing military campaigns. They arranged the creation of a Superintendency of the Imperial Encampment (*zongli xingying chu*) that double-checked the imperial route and encampment sites and determined the number of imperial bodyguards securing the imperial procession and bivouacs. Lower ranking officials inspected the roads, recorded place names and bridges, analysed the terrain and the distances between the encampments. This information was passed on to the Superintendency and the executive institutions responsible for construction work, armament and security patrols. They worked closely together with the governors-general and governors of the regions to be passed through and who were in charge of all preparations on the local level.[216] Each province furthermore disposed of a Central Office for the Management of Special Affairs (*banli chaiwu zongju*) that served to mobilize provincial resources for large imperial enterprises such as wars, inspection tours, or for disaster relief, and can be compared to the military supply and expenditure bureaus (*junxuju*).

It is impossible not to see the parallels to the preparations for military campaigns. Chang says that "the practices of touring and hunting overlapped and were not mutually exclusive".[217] The aspect of warfare ought to be added to these, too. The similarity also refers to particular details of transport: For one single inspection tour 18,000 carts were hired, 2,400 mules, 2,200 camels, 3,450 boats, up to 20,000 horses, as well as 300,000 boat towers, porters and construction personnel.[218] Officials on the lower levels of administration had to look after the stockpiling of rice, wheat, firewood or charcoal, and to purchase vegetables, meat and fish on a daily basis from the local markets. The omnipresent spectre of 'vicious elements' hoarding grain and of clerks and runners of the local *yamen*

[215] Chang, *Court on Horseback*, p. 116.
[216] Chang, *Court on Horseback*, p. 121.
[217] Chang, *Court on Horseback*, p. 138.
[218] Chang, *Court on Horseback*, p. 139; see also pp. 441–450.

extorting and expropriating money had also to be fought ruthlessly during these inspection tours.[219]

Yet not only in the matter of procuring supplies did the inspection tours strongly resemble the many dynastic wars. In an appendix to his book Michael Chang attempts to solve the question how much money the southern tours had cost, and whether they had contributed to the dynastic decline, just like the many wars that had depleted the government treasury. He asserts that, due to the absence of data, and because of unclear boundaries between the public treasury and the privy purse of the Imperial Household, it would be highly speculative to assess the concrete revenues and expenditure of the Qing state in the eighteenth century. It seems, in the end, that the (seemingly) inexhaustible revenue and the abundant reserves of more than 70 million silver *liang*—at least partially caused by the increased silver imports from the 1760s on—tempted the Qianlong emperor to spend more than he should have actually done.[220] Michael Chang carefully calculated the cost of a single inspection tour with the help of data from different sources, and reaches a total of just over 3 million *liang*,[221] which corresponds to the cost of the smaller early campaigns in Eastern Turkestan. Half of this sum, by the way, was extra-provincial expenditure accumulated as the result of rampant corruption in the salt voucher procedures. The share of the province of Jiangsu is officially reported as about 577,000 *liang* which in this case can be compared with the share of the war-waging province during a campaign. Tax reductions and debt annulments were proclaimed during the tour in order to reduce the fiscal burden on the districts and prefectures through which the imperial procession toured.[222] These sums should strictly speaking be added to the total cost of the enterprise because they reduced the revenue of future years as an indirect consequence of the tour.

The relation of 3 million *liang* for a prodigious inspection tour in 1786 to a four years long single war costing 60 million *liang* shows that the inspection tours were barely a factor in leading the dynasty in to bankruptcy,

219 A *yamen* was the seat of the local government in districts and prefectures and included administrative and jurisdictional institutions, shrines, as well as jails.

220 Lin Man-houng, *China Upside Down: Currency, Society, and Ideologies, 1808–1856* (Cambridge, Mass.: Harvard University Press, 2006), pp. 39–71, gives a concise overview over the importance and sources of the silver currency in Qing China.

221 Chang, *Court on Horseback*, p. 458.

222 Chang, *Court on Horseback*, p. 147.

although they ranked on the same level of cost intensity.[223] The question remains, which sources were tapped to finance the inspection tours, and, indirectly, the many officials underreporting expenditure and embezzling state funds. Besides the funds coming from the government on various levels, wealthy local merchants were urged by the local officials to contribute to the public cause, a phenomenon that could also be seen in the more regularized form of contribution campaigns during wars. A sizable amount of money from private sources was also used for the "beautification of local academies, temples, private residences and gardens", in order to impress the emperor.[224] The financing of the tour therefore can also be compared with that of military campaigns.

The regulations of the War Supplies and Expenditure Code include a number of administrative posts whose usefulness for a war is not immediately evident. This applies to the rules for the amount of salted-vegetable pay, the numbers of manservants and horses allowed civilian officials from central government institutions. Among those were academicians, readers-in-waiting, secretaries and clerks of the Grand Secretariat (*neige*), Grand Secretaries, archivists, editorial assistants, erudites, instructors, expositors-in-waiting and libationers (directors) of the Hanlin Academy and the Directorate of Education (*guozijian*); employees in the Ministries (*buyuan*); officials in the Censorate (*duchayuan*); Grand Ministers of the Imperial Household Department (*neidachen*); supervisors, recorders, mentors, admonishers and administrators from the Household of the Heir Apparent (*zhanshifu*), the princely establishments (*wangfu*) and the Court of the Imperial Clan (*zongrenfu*); members of the Directorate of Astronomy (*qintianjian*); commissioners, administrative assistants, medical secretaries and imperial physicians from the Imperial Academy of Medicine (*taiyiyuan*); personnel from the Court of Imperial Sacrifices (*taichangsi*), the Court of Imperial Entertainments (*guanglusi*), the Imperial Stud (*taipusi*), the Court of Judicial Review (*dalisi*), and the Court of State Ceremonial (*honglusi*); as well as commissioners of the Office of Transmission (*tongzhengshi*) and their assistants and registrars.[225]

---

223 If a hundred days' long tour cost 3 million *liang*, it cost about 1 million *liang* a month, which is roughly the same sum that the second Jinchuan campaign had cost in the same time frame, and with probably the same number of personnel involved. See Chang, *Court on Horseback*, pp. 441–450.

224 Chang, *Court on Horseback*, p. 463.

225 *Hubu junxu zeli*, juan 2, fol. 2a–3b. *Bingbu junxu zeli*, juan 1, fol. 6a–7a.

Some of these offices could certainly be of use during wars, for instance, censors that investigated cases of corruption, or physicians that looked after a member of the imperial family taking part in the campaign. A closer look at the institutions included in the lists shows that more or less all institutions and positions of the central government are mentioned. It might therefore be that the regulations of the *Junxu zeli* could also be applied for the organization and the auditing process of imperial inspection tours and not only for military campaigns. This assumption can be supported by the fact that the imperial canon on the southern tours, the *Nanxun shengdian*, does not include any regulations concerning organization and payment.[226]

Not only wars brought all actors within the empire together for concerted action, but also pompous official events like inspection tours and imperial hunts. During such events society was "involved at large":[227] Military and civilian officials, members of the central government and the local governments, soldiers of different branches of service, members of the local gentry and common people, peasants and urban residents, tradesmen and merchants, and, finally, the different ethnic groups of Manchus, Mongols, Chinese and native tribes. The state prodded the provincial bureaucracy, but also all other social groups, "into a heightened state of activity and readiness" as Michael Chang says, to achieve organizational discipline and a tighter integration of China's regions, as well as a higher level of general security.[228] Such an idealized picture of cooperation for the 'public cause' hides the fact that the bureaucracy as well as the common people and the merchant associations paid a high price for the military and ritual achievements of the dynasty.

The image of a successfully cooperating society that the Qianlong emperor created contrasts sharply with the social disintegration that began in the second half of the eighteenth century, with rampant corruption at the top and increasing banditry in the rural areas. Mobilization, financial as well as logistic, in the end meant that the government was able to pass on the responsibility of organizing and financing huge projects like a war down to the local bureaucracy, where it did not stop: It could be seen that the purchase and transport of grain was ever more often laid into the hands of private entrepreneurs, that part of the organization of the grain

---

226 *Nanxun shengdian* (Siku quanshu edition).
227 Dai, "Military finance", p. 307.
228 Chang, *Court on Horseback*, p. 141.

supply was done by unpaid candidates for office, and that approximately one tenth of a large campaign was financed by salt merchant associations. More and more, even the responsibility to fight was laid into the hands of the local gentry, as Philipp Kuhn has shown.[229] He explains that the *tuan* militia with their gentry leadership were brought into the formal structure of the local government and began to function as an official sub-district administrative organ. Power was in this way shifted from a "relatively uncontrollable and dangerous group (the [exploitative and corrupt] clerks and runners) to a relatively sympathetic and predictable group (the gentry)", and the gentry, with its near-official status, gradually took over local administration.

These developments were part of the immense expansion of the Qing empire between 1700 and 1800. The bureaucratic absolutism of the Qing with its profusion of jurisdictional regulations for all aspects of statecraft reached maturity during the last decades of the Qianlong reign. It enabled the Manchu dynasty to mobilize manpower, existing or newly established infrastructure, and various funds to conquer the lands of formidable enemies like the Dzungars or the people of Jinchuan. Yet unlike in Europe, where the cost of warfare was clearly visible as estates or parliaments demanded a submission of accounts, the cost of the Qianlong emperor's wars were, in spite of all seeming administrative rationality, hidden in social dislocations. These were to ruin China even before the great disasters of internal turmoil and external threat took shape in the first half of the nineteenth century.

[229] Philipp Kuhn, *Rebellion and its Enemies in Late Imperial China: Militarization and Social Structure, 1796–1864* (Cambridge, Mass.: Harvard University Press, 1970), pp. 212, 214.

APPENDIX ONE

NAMES AND DESIGNATIONS OF QING EMPERORS, AND REMARKS TO THE CALENDAR

Posthumous temple name	Reign motto/period	Duration of reign*
清世祖 Qing Shizu	順治 Shunzhi (SZ)	1644–1661
清聖祖 Qing Shengzu	康熙 Kangxi (KX)	1662–1722
清世宗 Qing Shizu	雍正 Yongzheng (YZ)	1723–1735
清高宗 Qing Gaozong	乾隆 Qianlong (QL)	1736–1795
清仁宗 Qing Renzong	嘉慶 Jiaqing (JQ)	1796–1820
清宣宗 Qing Xuanzong	道光 Daoguang (DG)	1821–1850
清文宗 Qing Wenzong	咸豐 Xianfeng (XF)	1851–1861
清穆宗 Qing Muzong	同治 Tongzhi (TZ)	1862–1874
清德宗 Qing Dezong	光緒 Guangxu (GX)	1875–1908
(清末帝 The Last Emperor)	宣統 Xuantong (XT)	1909–1912

* The years indicate the time in which the reign mottoes were valid. The respective emperors' reign begins immediately after the death or abdication of his predecessor.

It should be noted that these dates do not take into account the time difference between the Chinese lunar year and the Western year. The Shunzhi reign, for instance, actually began on February 8, 1644 (first day of the first lunar month of the Chinese calendar), and ended on February 17, 1662 (last day of the twelfth lunar month), which was the 29th day of the twelfth month of the 18th year of the Shunzhi reign. The day after was the first day of the first month of the first year of the Kangxi reign, or, according to the Western calendar, February 18, 1662.

Dates in primary sources are always given according to the pattern 'Reign motto and year/month/day' (for instance, QL 40/9/jiaxu, meaning: 40th year of the Qianlong reign, 9th month, day with the cyclical combination *jiaxu*). The Chinese calendar makes use of sixty combinations of two sets of characters (the so-called ten Celestial Stems and twelve Terrestrial Branches), resulting in a continuous cycle of sixty fixed combinations of signs that are used for days, and also for years. Alternatively, days are also numbered from one to 29 or 30. The Chinese calendar also knows intercalary months (*runyue*, full thirteenth months) that are marked in the footnotes with the letter 'r' (for instance, QL 13/r7/jisi, meaning: 13th year of the Qianlong reign, intercalary 7th month, day with the cyclical combination *jisi*).

APPENDIX TWO

LIST OF QING PERIOD WARS

Shunzhi Reign

1644–1661	initial conquest of Ming China and suppression of the Southern Ming
1651–1663	invasions by Zheng Chenggong 鄭成功 (Coxinga)
1652, 1654	conflicts with Russia about Ujala (Amursk) and Khumarsk (Huma, Heilongjiang)

Kangxi Reign

1663	rebellion of Zhou Yu 周玉 in Guangdong
1663–1664	rebellion of the remaining adherents of Li Zicheng 李自成 in Fujian
1664–1669	rebellions of native peoples in Guizhou and Yunnan
1671	peasant rebellion in Shanxi
1673–1681	war against the Three Feudatories of Wu Sangui 吳三桂, Shang Kexi 尙可喜 (resp. his son Shang Zhixin 尙之信) and Geng Jingzhong 耿精忠, and that of Wang Fuchen 王輔臣 and Sun Yanling 孫延齡
1674	uprising of Huang Jinlong 黃金龍 and He Shirong 何士戎 in Hubei
1675	war against the Chahar Mongol prince Burni
1676–1683	war against the Zheng 'dynasty' under Zheng Jing 鄭經, conquest of Taiwan
1677	rebellion of the Ming prince Zhu Tongchang 朱統錩 in Fujian
	rebellion of Yang Yutai 楊玉泰 in Jiangxi
1677–1696	war against the Dzungars under Galdan
1680	rebellion of Yang Qilong 楊起隆 in Shanxi
1685–1686	campaign against the Cossacks of Albazin (Yaksa)
1688	rebellion of Xia Fenglong 夏逢隆 in Hubei
1698	suppression of the last adherents of Wu Sangui
1700	conquest of Dartsedo (Eastern Tibet)
1702, 1704	rebellion of Miao tribes in Hunan and Yao tribes in Guangdong
1705	rebellion of Li Tianij 李天極 and Zhu Liufei 朱六非 in Yunnan
1707	rebellion of Miao tribes in Guizhou
1708	uprising of the Heaven-and-Earth Society (*tiandihui* 天地會) in Zhejiang
1710	rebellion of Chen Wuxian 陳五顯 in Fujian
1715–1721	war against the Dzungars under Tsewang Rabdan
1717–1718	conquest of Tibet
1721–1724	conquest of Qinghai
1721	rebellion of Zhu Yigui 朱一貴 in Taiwan

Yongzheng Reign

1726–1730	rebellion of Miao tribes in Yunnan and Guizhou
1727	campaign against the regent of Tibet, Nga-phod-pa rDo-rje rGyal-po
1728	rebellion of the native ruler Yan Guangse 顏光色 in Guangxi
1730–1732	campaign against the Dzungars under Galdan Tsering
1732	rebellion of Lin Wuli 林武力 and Wu Fusheng 吳福生 in Taiwan
1732–1735	rebellion of Miao tribes in Guizhou, Yunnan and Hunan

Qianlong Reign

1738–1741	rebellions of Miao, Yao and Tong tribes in Guizhou and Hunan
1745	conquest of Djandui (Garzê, Eastern Tibet)
1748–1749	first war against the native ruler of Chu-chen (Greater Jinchuan, Eastern Tibet)
1750	campaign against the Tibetan regent Gyur-med rNam-rgyal
1752	conquest of Dzagu (Eastern Tibet)
1755	first war against the Dzungars
1756–1757	second war against the Dzungars under Amursana
1758–1759	campaign against the Greater and Lesser Khojas in Yarkant and Kašgar
1765	Muslim insurgency in Uchturpan (Uš)
1766–1769	Myanmar (Burma) campaigns
1770	Miao rebellion in Guizhou
1770	rebellion of the sectarian leaders Chen Zongbao 陳宗寶 and Zheng Chun 鄭純 in Taiwan
1771–1776	second Jinchuan war against Chu-chen (Greater Jinchuan) and bTsan-lha (Lesser Jinchuan)
1774	rebellion of Wang Lun's 王倫 Pure Water Sect (*qingshuijiao* 清水教) in Shandong
1781	Muslim insurgency under Su Sishisan 蘇四十三 in Lanzhou
1781	rebellion of Hu Fannian 胡范年 in Sichuan
1784	Muslim insurgency under Tian Wu 田五 and Ma Sigui 馬四圭 in Fort Shifeng, Gansu
1786	rebellion of Lin Shuangwen 林爽文 in Taiwan
1788	Annam campaign
1788–1792	two campaigns in Tibet against the Gurkhas from Nepal
1795	Miao rebellion in Hunan and Guizhou

Jiaqing Reign

1796–1804	rebellion of the White Lotus Sect (*bailianjiao* 白蓮教)
1797	rebellion of Buyi tribes under the female chieftain Wang Nangxian 王囊仙 in Guizhou
1802	rebellion of Lisu tribes in Yunnan
1803–1809	rebellion of Cai Qian 蔡牽 in Fujian, invasion of Taiwan

1813 sectarian uprising of the Heavenly Order sect (*tianlijiao* 天理教) under Li Wencheng 李文成 and Lin Qing 林清 in north China

1815 rebellion of Ziya' ad-dīn in Kašgar

1817–1821 rebellions of Yi tribes in Yunnan

1820 invasion by Jahāngīr from Kokand in Eastern Turkestan

Daoguang Reign

1822 rebellion of Tibetan tribes in Qinghai

1826–1827 second invasion of Jahāngīr in Eastern Turkestan

1830 invasion by Yūsup from Kokand in Eastern Turkestan

1832, 1836 rebellion of Yao tribes in Hunan and Guangdong

1832 sectarian uprising of the Heaven-and-Earth Society under Zhang Bing 張丙 and Chen Ban 陳辦 in Taiwan

1833 rebellion of Yi tribes in Sichuan

1835 sectarian uprising of the Former Heaven Sect (*xiantianjiao* 先天教) under Cao Shun 曹順 in Shanxi

1838 rebellion of Mu Jixian 穆繼賢 in Sichuan

1839–1842 First Opium War

1842 rebellion of Zhong Renjie 鍾人傑 in Hubei

1843 rebellion of Zeng Ruzhu 曾如炷 in Hunan

1844 rebellion of Yang Dapeng 楊大鵬 in Hunan

uprising of the Heaven-and-Earth Society under Hong Xie 洪協 in Taiwan

1846 rebellion of Wang Zongxian 王宗獻 in Hunan

1847, 1849 rebellion of the Green Lotus Sect (*qinglianjiao* 青蓮教) under Lei Zaihao 雷再浩 and Li Yuanfa 李沅發 in Hunan

1848 rebellion of Jin Derun 金德潤 in Jiangsu

Xianfeng Reign

1850–1864 war against the Heavenly Kingdom (*taiping tianguo* 太平天國) of Hong Xiuquan 洪秀全

1852 uprising of the Heaven-and-Earth Society under Wu Lingyun 吳凌雲 and Zhu Hongying 朱洪英 in Guangxi

1853 uprising of the Small Sword Society (*xiaodaohui* 小刀會) under Huang Wei 黃威 in Fujian and under Liu Lichuan 劉麗川 Shanghai

uprising of the Red Cash Society (*hongqianhui* 紅錢會) under Lin Jun 林俊 and Chen Hu 陳湖 in Fujian

uprising of the Heaven-and-Earth Society under Zhu Jiutao 朱九濤 in Hunan

1854 uprising of the Heaven-and-Earth Society under Chen Kai 陳開 in Guangdong

rebellion of Buyi tribes in Guizhou

sectarian uprising of Yang Longxi 楊隆喜 and Shu Caifeng 舒裁縫 in Guizhou

1855	Gurkha invasion in Tibet
	rebellion of the Red Signal Army (*honghaojun* 紅號軍) of the Flower Lantern Sect (*huadengjiao* 花燈教) under Wu Qulao 吳劬勞 and Chen Fulin 陳福林 in Guizhou
	uprising of Miao tribes under Zhang Xiumei 張秀眉 in Guizhou
1856	rebellion of Muslims and Yi tribes in Yunnan
1857–1859	Second Opium War (Arrow War), occupation of Beijing
1858–1859	new rebellion of Zhang Xiumei in Guizhou
	rebellion of the White and Yellow Signal armies of the Flower Lantern Sect under Liu Yishun 劉義順 and He Guan 何冠 in Guizhou
	rebellion of the 'White Banner' (*baiqi* 白旗) under the Muslim leaders Zhang Lingxiang 張凌翔 and Ma Hetu 馬河圖 in Guizhou
1860	rebellion of Lan Chaoding 藺朝鼎 and Li Yonghe 李永和 in Sichuan
1861	uprising of the Eight Hexagrams Sect (*baguajiao* 八卦教) in Shandong
	uprising of the Gold Cash Society (*jinqianhui* 金錢會) in Zhejiang

Tongzhi Reign

1862	rebellion of Dai Wansheng 戴萬生 in Taiwan
	Muslim rebellion in Shaanxi
	rebellion of Tong tribes in Guizhou
1863	rebellion of Miao Peilin 苗沛霖 in Anhui
	rebellion of Song Jingshi 宋景詩 in Shandong
	Muslim rebellions in Shaanxi, Gansu and Qinghai
1863–1868	Nian rebellion
1864	large-scale rebellion in Xinjiang under Ma Long 馬隆
	rebellion of Lan Dashun 藺大順 in Shaanxi
	rebellion of Tong tribes in Guizhou
1866, 1868	war against the Signal Armies (*haojun* 號軍) in Guizhou
1867–1868	third rebellion of Zhang Xiumei in Guizhou
	Miao rebellion in Guizhou
	Muslim rebellion under Du Wenxiu 杜文秀 in Yunnan
1869	Muslim rebellion under Ma Shilong 馬仕龍 in Gansu
1871	Muslim rebellion in Yunnan
1871–1873	Muslim rebellion under Ma Zhan'ao 馬占鰲 in Gansu

Guangxu Reign

1875	rebellion of Song Sanhao 宋三好 in Liaodong
	invasion of Vietnam in pursuit of the rebel Huang Chongying 黃崇英
1876	secession of Yakub Beg
1883–1885	Sino-French war over Tongking (northern Vietnam)
1888	armed conflict with Great Britain over Tibet

1889	rebellion of Shi Jiuduan 施九段 in Taiwan
1890	rebellion of Yu Dongchen 余棟臣 in Sichuan
1891	rebellion of the Golden Elixir Sect (*jindanjiao* 金丹教) under Yang Yuechun 楊悅春 and Li Guozhen 李國珍 in Rehe
1894–1895	first Sino-Japanese war
1895	Muslim rebellion under Ma Yonglin 馬永琳 in Gansu
1900	attempted putsch of the Revive-China Society (*xingzhonghui* 興中會) under Sun Yat-sen in Huizhou, Guangdong
	Boxer rebellion and invasion of Beijing by the Eight-Nation Alliance
1904	rebellion under Wang Heshun 王和順 in Guangxi
	armed conflict with Great Britain over Tibet
	rebellion in Batang, Eastern Tibet, against the abolition of native rule
1906–1907	attempted coups by Sun Yat-sen's United Alliance Society (*tongmenghui* 同盟會)

Note: Many of the rebellions in the list above did surpass the local level, especially those uprisings that were directed against the exploitation by the local gentry, and not against the Qing dynasty.

Source: Fu Zhongxia 傅仲俠 et al., eds. *Zhongguo junshi shi* 中國軍事史, *Fujuan: Lidai zhanzheng nianbiao* 附卷：歷代戰爭年表, vol. 2 (Beijing: Jiefangjun chubanshe, 1986).

APPENDIX THREE

WEIGHTS AND MEASURES

Weight Units

1 *jin* = 596.8 g	兩	錢	分	厘	毫	絲	忽(微)	撮
兩 *liang* 'tael'	1	0.1	0.01	0.001	0.0001	10^{-5}	10^{-6}	10^{-7}
錢 *qian* 'mace'	10	1	0.1	0.01	0.001	0.0001	10^{-5}	10^{-6}
分 *fen* 'candareen'	100	10	1	0.1	0.01	0.001	0.0001	10^{-5}
厘 *li* 'cash'	1,000	100	10	1	0.1	0.01	0.001	0.0001
毫 *hao*	10,000	1,000	100	10	1	0.1	0.01	0.001
絲 *si*	10^5	10,000	1,000	100	10	1	0.1	0.01
忽 *cong* (微 *wei*)	10^6	10^5	10,000	1,000	100	10	1	0.1
撮 *cuo*	10^7	10^6	10^5	10,000	1,000	100	10	1

兩 *liang*	斤 *jin*	兩 *liang*	斤 *jin*	兩 *liang*	斤 *jin*	兩 *liang*	斤 *jin*
1	0.0625	5	0.3125	9	0.5625	13	0.8125
2	0.125	6	0.375	10	0.625	14	0.875
3	0.1875	7	0.4375	11	0.6875	15	0.9375
4	0.25	8	0.5	12	0.75	16	1

1 *qian* = 0.1 *liang* = 0.00625 *jin*
1 石 *shi* 'picul' = 100 *jin* 'catties'
1 *liang* = 37.4–37.8 g

Volume Units

1 *sheng* = 1035 ml	石	斛	斗	升	合	勺	撮
石 *dan*	1	0.5	0.1	0.01	0.001	0.0001	10^{-5}
斛 *hu*	2	1	0.5	0.05	0.005	0.0005	5×10^{-4}
斗 *dou*	10	5	1	0.1	0.01	0.001	0.0001
升 *sheng*	100	50	10	1	0.1	0.01	0.001
合 *ge*	1,000	500	100	10	1	0.1	0.01
勺 *shao* (抄 *chao*)	10,000	5,000	1,000	100	10	1	0.1
撮 *cuo*	100,000	50,000	10,000	1,000	100	10	1

Length and Area Units

1 *chi* = 32 cm	丈	尺	寸	分	厘
丈 *zhang*	1	0.1	0.01	0.001	0.0001
尺 *chi*	10	1	0.1	0.01	0.001
寸 *cun*	100	10	1	0.1	0.01
分 *fen*	1000	100	10	1	0.1
厘 *li* (微 *wei*)	10,000	1000	100	100	1

1 *li* = 480 m
1 *mu* = 614.4 m^2

Source: Qiu Guangming, Qiu Long and Yang Ping. *Zhongguo kexue jishu shi: Duliangheng juan* (Beijing: Kexue chubanshe, 2001), p. 447.

BIBLIOGRAPHY

1. Primary Sources

Authors and Editors

Amiot, Joseph. "Lettre du P. Amiot, Missionaire de la Chine, sur la Réduction des Miao-tsé, en 1775" [Letter by Father Amiot, Missionary in China, on the pacification of the Miaozi, scil. the Jinchuan people, in 1775; written on Sep 12, 1776], in *Mémoires Concernant l'Histoire, les Sciences, les Arts, les Moers, les Usages &c. des Chinois, par les Missionaires de Pékin* [The Beijing missionaries' memoirs on the history, sciences, arts, customs and habits of the Chinese] (Paris: Nyon, 1778), vol. 3, pp. 387–422.

Cheng Muheng 程穆衡. *Jinchuan jilüe* 金川紀略 [Concise account of the (second) Jinchuan war], in *Zhongguo yeshi jicheng* 中國野史集成 [Collection of unofficial histories of China] (Chengdu: Ba-Shu shushe, 1993), vol. 40, pp. 40–71.

Dalrymple, Campbell. *A Military Essay: Containing Reflections on the Raising, Arming, Cloathing, and Discipline of the British Infantry and Cavalry* (London: Wilson, 1761; also to be found on http://books.google.de/books/about/A_military_essay.html?id=xyNEAAAAYAAJ&redir_esc=y).

Ding Baozhen 丁寶楨, ed. *Sichuan yanfa zhi* 四川鹽法志 [The salt administration of Sichuan] (print from the Guangxu reign; electronic version [data bank] used via licence by Staatsbibliothek Berlin of Zhongguo jiben guji ku 中國基本古籍庫, ed. by Beijing Airusheng shuzihua jishu yanjiu zhongxin 北京爱如生数字化技术研究中心, Beijing).

Faššan (Fashishan) 法式善. *Taolu zalu* 陶廬雜錄 [Miscellaneous notes from the 'Potter's Hut'], in *Qingdai shiliao biji congkan* 清代史料筆記叢刊 (Beijing: Zhonghua shuju, 1959).

Godfrey, Michael. *A Short Account of the Bank of England* (London: John Whitlock, 1695; republ. in *Central Banking in History*, ed. by Michael Collins, Aldershot: Hants, and Brookfield, Vt.: Elgar, 1993; also published in Early English Books Online, http://eebo.chad wyck.com/search/full_rec?SOURCE=pgimages.cfg&ACTION=ByID&ID=V36202).

Gui Fu 桂馥. *Zhapu* 札樸 [Rough notes] (print by Xiaoli shanfang 小李山房, 1813; electronic version [data bank] used via licence by Staatsbibliothek Berlin of Zhongguo jiben guji ku 中國基本古籍庫, ed. by Beijing Airusheng shuzihua jishu yanjiu zhongxin 北京爱如生数字化技术研究中心, Beijing).

Huang Liuhong 黃六鴻. *Fuhui quanshu* 福惠全書 [All about wealth and benefits] (print by Jinling lianxi shuwu 金陵濂溪書屋, 1696; electronic version [data bank] used via licence by Staatsbibliothek Berlin of Zhongguo jiben guji ku 中國基本古籍庫, ed. by Beijing Airusheng shuzihua jishu yanjiu zhongxin 北京爱如生数字化技术研究中心, Beijing).

Jiao Xu 焦勖. *Huogong qieyao* 火攻挈要 [All important matters of gunfight], in *Haishanxianguan congshu* 海山仙館叢書 (also to be found in *Zhongguo kexue jishu dianji tonghui: Jishu juan* 中國科學技術典籍通彙：技術卷, Zhengzhou: Henan jiaoyu chubanshe, 1994, vol. 5, pp. 1265–1321).

Li Xinheng 李心衡. *Jinchuan suoji* 金川瑣記 [Petty notes on Jinchuan], in *Congshu jicheng* 叢書集成, vol. 3199.

Liu Yumo 劉餘謨. "Kenhuang xingtun shu 墾荒興屯疏" [On the reclamation of land and the founding of military colonies], in *Huangchao jingshi wenbian* 皇朝經世文編, juan 34 (published in *Jindai Zhongguo shiliao congkan* 近代中國史料叢刊, series 74, no. 731, vol. 3, pp. 1234–1235).

Nayancheng 那彥成. *A Wenchenggong nianpu* 阿文成公年譜 [A biography of A Wenchenggong, i.e. Agui] (print from 1813; electronic version [data bank] used via licence by Staatsbibliothek Berlin of Zhongguo jiben guji ku 中國基本古籍庫, ed. by Beijing Airusheng shuzihua jishu yanjiu zhongxin 北京爱如生数字化技术研究中心, Beijing).

Shen Shucheng 沈書城. *Zeli bianlan* 則例便覽 [Convenient overview of administrative precedent rules] (print from 1791; electronic version [data bank] used via licence by Staatsbibliothek Berlin of Zhongguo jiben guji ku 中國基本古籍庫, ed. by Beijing Airusheng shuzihua jishu yanjiu zhongxin 北京爱如生数字化技术研究中心, Beijing).

Shi Menglan 史夢蘭. *Zhiyuan bitan* 止園筆談 [Brush discussions from the 'Garden of Recreation'] (print from 1878; electronic version [data bank] used via licence by Staatsbibliothek Berlin of Zhongguo jiben guji ku 中國基本古籍庫, ed. by Beijing Airusheng shuzihua jishu yanjiu zhongxin 北京爱如生数字化技术研究中心, Beijing).

Sun Yuanhua 孫元化. *Xifa shenji* 西法神機 [Miraculous weapons of the West], in *Zhongguo kexue jishu dianji tonghui: Jishu juan* 中國科學技術典籍通彙：技術卷 (Zhengzhou: Henan jiaoyu chubanshe, 1994), vol. 5, pp. 1233–1264.

Wang Chang 王昶. *Shujiao jiwen* 蜀徼紀聞 [An account from the borderlands of Shu, i.e. Sichuan], in *Xiaofanghuzhai yudi congchao* 小方壺齋輿地叢鈔, part 8, vol. 43.

Wei Yuan 魏源. *Shengwuji* 聖武記 [The sacred wars (of Our Dynasty)], in *Sibu beiyao* 四部備要.

Xu Guoying 許國英. *Jinchuan yaoji zhi* 金川妖姬志 [The seductress of Jinchuan], in *Man-Qing yeshi sanbian* 滿清野史三編 [Unofficial histories of the Manchu-Qing, third series], No. 19 (republ. in *Zhongguo yeshi jicheng* 中國野史集成 [Collection of unofficial histories of China] [Chengdu: Ba-Shu shushe, 1993], vol. 40, pp. 106–111).

Xu Ke 徐珂. *Qingbai leichao* 清稗類鈔 [Classified anecdotes of Qing history] (Shanghai: Shangwu yinshuguan, 1917; electronic version used of Kaifang wenxue 開放文學, http://www.open-lit.com).

Zhao Shizhen 趙士楨. *Shenqipu* 神器譜 [Miraculous weapons], in *Xuanlantang congshu* 玄覽堂叢書 (also publ. in *Zhongguo kexue jishu dianji tonghui: Jishu juan* 中國科學技術典籍通彙：技術卷, Zhengzhou: Henan jiaoyu chubanshe, 1994, vol. 5, pp. 633–656).

Zhao Yi 趙翼. *Pingding Liang Jinchuan shulüe* 平定兩金川述略 [A short account of the war against the Two Jinchuans, i.e. the second Jinchuan war], in *Xiaofanghuzhai yudi congchao* 小方壺齋輿地叢鈔, part 8, no. 43 (also publ. as part of *Huangchao wugong jisheng* 皇朝武功紀盛, in *Congshu jicheng* 叢書集成, no. 3999).

——. *Yanpu zaji* 簷曝雜記 [Miscellaneous notes from the 'Sunning on the Eaves Studio'] in *Qingdai shiliao biji congkan* 清代史料筆記叢刊 (Beijing: Zhonghua shuju, 1959).

Zhaolian 昭槤 (Jaoliyan). *Xiaoting zalu* 嘯亭雜錄 [Miscellaneous records of the 'Whistling Pavilion'], in *Qingdai shiliao biji congkan* 清代史料筆記叢刊 (Beijing: Zhonghua shuju, 1980).

Zhou Xun 周詢. *Shuhai congtan* 蜀海叢談 [Collected notes on the ocean-vast affairs in Shu, i.e. Sichuan], in *Jindai Zhongguo shiliao congkan* 近代中國史料叢刊, series 1, no. 7.

Corporate Authors

"Bank of England Act 1694" (http://www.legislation.gov.uk/aep/WillandMar/5-6/20/introduction).

Baqi tongzhi 八旗通志 [Comprehensive treatises on the Eight Banners] (Changchun: Jilin wenshi chubanshe, 2002).

Bingbu junxu zeli 兵部軍需則例, see *Junxu zeli* 軍需則例.

Da-Qing huidian 大清會典 [Collected statutes of the Great Qing] in *Siku quanshu* 四庫全書 (Guangxu edition also publ. by Taibei: Zhongwen shuju, 1963).

Da-Qing lichao shilu Sichuan shiliao 大清歷朝實錄四川史料 [Documents about Sichuan from the Veritable Records of the Great Qing through all reign periods], ed. by Wang Gang 王綱 (Chengdu: Dianzi keji daxue chubanshe, 1991; a planned second volume has never been published).

(*Jiaqing chongxiu*) *Da-Qing yitong zhi* (嘉慶重修) 大清一統志 [The imperial geography of the Great Qing, revised Jiaqing version] in *Sibu congkan xubian* 四部叢刊續編.

Gaozong shilu 高宗實錄, see *Qingshilu* 清實錄.

Gongbu junqi zeli 工部軍器則例 [Regulations of the Ministry of Works for weapons and military equipment], comp. by Dong Gao 董誥 et al. (print from 1812; electronic version [data bank] used via licenceby Staatsbibliothek Berlin of Zhongguo jiben guji ku 中國基本古籍庫, ed. by Beijing Airusheng shuzihua jishu yanjiu zhongxin 北京爱如生数字化技术研究中心, Beijing).

Gongbu junxu zeli 工部軍需則例, see *Junxu zeli* 軍需則例.

Gongbu xuzeng zeli 工部續增則例 [Extended regulations of the Ministry of Works] (Jiaqing 22 version), in *Gugong zhenben congkan* 故宮珍本叢刊, vol. 294–298, as *Gongbu zeli san zhong* 工部則例三種類 [Three editions of the "Regulations of the Ministry of Works"] (Haikou: Hainan chubanshe, 2000).

Gongbu zeli 工部則例 [Regulations of the Ministry of Works] (Qianlong 14 version), in *Zhongguo gudai jiangzuo ziliao congkan* 中國古代匠作資料叢刊, as *Gongbu zeli zheng-xubian* 工部則例正續編 [Regulations of the Ministry of Works, first version and extensions] (Beijing: Bejing tushuguan chubanshe, 1997).

Gongbu zeli 工部則例 [Regulations of the Ministry of Works] (Qianlong 58 version), in *Zhongguo gudai jiangzuo ziliao congkan* 中國古代匠作資料叢刊, as *Gongbu zeli zheng-xubian* 工部則例正續編 [Regulations of the Ministry of Works, first version and extensions] (Beijing: Bejing tushuguan chubanshe, 1997).

Gongbu zeli 工部則例 [Regulations of the Ministry of Works] (Jiaqing 20 version), in *Gugong zhenben congkan* 故宮珍本叢刊, vol. 294–298, as *Gongbu zeli san zhong* 工部則例三種類 [Three editions of the "Regulations of the Ministry of Works"] (Haikou: Hainan chubanshe, 2000).

Gongbu zeli 工部則例 [Regulations of the Ministry of Works] (Guangxu 10 version), in *Gugong zhenben congkan* 故宮珍本叢刊, vol. 294–298, as *Gongbu zeli san zhong* 工部則例三種類 [Three editions of the "Regulations of the Ministry of Works"] (Haikou: Hainan chubanshe, 2000).

Guochao qixian leizheng 國朝耆獻類徵 [Topically arranged (biographies) of the honoured of Our Dynasty] (Guangxu version, Nanjing: Jiangsu Guangling guji keyinshe, 1990).

Huangchao jingshi wenbian 皇朝經世文編 [Collected writings on statecraft of Our August Dynasty], ed. by He Changling 賀長齡 (print from 1886; electronic version [data bank] used via licence by Staatsbibliothek Berlin of Zhongguo jiben guji ku 中國基本古籍庫, ed. by Beijing Airusheng shuzihua jishu yanjiu zhongxin 北京爱如生数字化技术研究中心, Beijing; also published in *Jindai Zhongguo shiliao congkan* 近代中國史料叢刊, series 74, no. 731).

Huangchao wenxian tongkao 皇朝文獻通考 [General history of institutions and critical examination of documents and studies of Our August Dynasty] (Qianlong 52), in *Shitong* 十通, as *Qingchao wenxian tongkao* 清朝文獻通考 (Hangzhou: Zhejiang guji chubanshe, 2000).

Huang-Qing zhigong tu 皇清職貢圖 [Illustrated account of the official tributary (nations) of the Imperial Qing], in *Siku quanshu* 四庫全書.

Hubu junxu zeli 戶部軍需則例, see *Junxu zeli* 軍需則例.

Jinchuan an 金川案 [Files from the Jinchuan wars], ed. by Zhongguo shehui kexue yuan minzu yanjiusuo 中国社會科学院民族研究所 and Sichuan shaoshu minyu shehui lishi diaocha zu 四川少數民族社會歷史調查組 (No place, 1963; also publ. as *Jinchuan an: Jinchuan liu zhong he ke* 金川案, 金川六种合刻 [Jinchuan files: Six books on Jinchuan, combined printing], ed. by Xizang shehui kexue yuan Xizangxue hanwen wenxian bianji shi 西藏社會科學院西藏學漢文文獻編輯室, Beijing: Zhongguo zangxue chubanshe, 1994).

Jinchuan dang 金川檔 [Archived documents of the (second) Jinchuan war], in *Guoli gugong bowuyuan diancang zhuan dang'an ji fanglüe congbian* 國立故宮博物院典藏專檔案暨方略叢編 (Taibei: Shenxiangting qiye, 2007).

Junjichu lufu zouzhe 軍機處錄副奏摺 [Document copies of the State Council] (Microfilms stored by Zhongguo di yi lishi dang'anguan 中國第一歷史檔案館 [First Historical Archives] in Beijing).

Junqi zeli 軍器則例 [Regulations for weapons and military equipment] (Qianlong 56 version), in *Gugong zhenben congkan* 故宮珍本叢刊 (Haikou: Hainan chubanshe, 2000), vol. 293.

Junqi zeli 軍器則例 [Regulations for weapons and military equipment] (Jiaqing 9 version) in *Gugong zhenben congkan* 故宮珍本叢刊 (Haikou: Hainan chubanshe, 2000), vol. 293.

Junqi zeli 軍器則例 [Regulations for weapons and military equipment] (Jiaqing 21 version) in *Xuxiu siku quanshu* 續修四庫全書, vol. 857.

Junxu zeli 軍需則例 [Regulations for war expenditure], in *Xuxiu siku quanshu* 續修四庫全書, vol. 857.

Nanxun shengdian 南巡盛典 [Imperial favours granted during the southern inspection tours], in *Siku quanshu* 四庫全書.

Pingding Jinchuan fanglüe 平定金川方略 [Military annals of the war against Jinchuan, i.e. the first Jinchuan war], in *Siku quanshu* 四庫全書 (also publ. by Xizang shehui kexue yuan 西藏社會科學院 in *Xizangxue Hanwen wenxian kuike* 西藏學漢文文獻彙刻, series 1, Beijing: Quanguo tushuguan wenxian suowei fuzhi zhongxin, 1991).

Pingding Liang Jinchuan fanglüe 平定兩金川方略 [Military annals of the war against the Two Jinchuans, i.e. the second Jinchuan war], in *Siku quanshu* 四庫全書 (also publ. by Xizang shehui kexue yuan 西藏社會科學院 in *Xizangxue Hanwen wenxian huike* 西藏學漢文文獻彙刻, series 1, Beijing: Quanguo tushuguan wenxian suowei fuzhi zhongxin, 1991; latest publ. in *Guoli gugong bowuyuan diancang zhuan dang'an ji fanglüe congbian* 國立故宮博物院典藏專檔案暨方略叢編, Taibei: Shenxiangting qiye, 2007).

Pingding Liang Jinchuan junxu li'an 平定两金川军需例案 [Precedents for war expenditure of the war against the Two Jinchuans], comp. by Zheng Qishan 鄭棲山, ed. by Xizang shehui kexue yuan 西藏社會科學院, in *Xizangxue Hanwen wenxian huike* 西藏學漢文文獻彙刻, series 2, Beijing: Quanguo tushuguan wenxian suowei fuzhi zhongxin, 1991).

Pingding sanni fanglüe 平定三逆方略 [Military annals of the war against the Three Feudatories], in *Siku quanshu* 四庫全書.

Pingding Zhungar fanglüe 平定準噶爾方略 [Military annals of the war against the Dzungars], in *Siku quanshu* 四庫全書.

(*Yongzheng*) *Sichuan tongzhi* (雍正) 四川通志 [Gazetteer of the province of Sichuan, Yongzheng version], in *Siku quanshu* 四庫全書.

(*Jiaqing*) *Sichuan tongzhi* (嘉慶) 四川通志 [Gazetteer of the province of Sichuan, Jiaqing version] (Chengdu: Ba-Shu shushe, 1984).

(*Tongzhi*) *Suzhou fu zhi* (同治) 蘇州府志 [Gazetteer of the prefecture of Suzhou, Tongzhi version] (Nanjing: Jiangsu guji chubanshe, 1991).

Qingguoshi 清國史 [History of the Qing (in biographies)], ed. by Wu Ge 吳格 (Jiayetang 嘉業堂 manuscript; Beijing: Zhonghua shuju, 1993).

Qingshi liezhuan 清史列傳 [Biographies of the Qing history], ed. by Wang Zhonghan 王鍾翰 (Beijing: Zhonghua shuju, 1987).

Qingshigao 清史稿 [Outline of (an official) history of the Qing dynasty], comp. by Ke Shaomin 柯劭忞 (Beijing: Zhonghua shuju, 1977).

Qingshilu 清實錄, i.e. *Da-Qing lichao shilu* 大清歷朝實錄 [The veritable records of successive reigns of the Qing dynasty]: *Shizu shilu* 世祖實錄 [The veritable records of Emperor Shizu (the Shunzhi emperor)], *Shizong shilu* 世宗實錄 [The veritable records of Emperor Shizong (the Yongzheng emperor)], *Gaozong shilu* 高宗實錄 [The veritable records of Emperor Gaozong (the Qianlong emperor)], *Renzong shilu* 仁宗實錄 [The veritable records of Emperor Renzong (the Jiaqing emperor)] (Taibei: Huawen shuju, 1964).

Qingshilu Zangzu shiliao 清實錄藏族史料 [Historiographical sources about the Tibetans from the Veritable Records], ed. by Gu Zucheng 顧祖成 et al. (Beijing: Xizang renmin chubanshe, 1982).

Zhongguo yanzheng yange shi: Changlu 中國鹽政沿革史：長蘆 [A history of salt administration in China: Changlu], ed. by Yanwushu 鹽務署 [Salt administration bureau, 1914], in *Jindai Zhongguo shiliao congkan* 近代中國史料叢刊, series 64, no. 637.

2. Secondary Literature

Authors in Western Languages

Ágoston, Gábor. *Guns for the Sultan: Military Power and the Weapons Industry in the Ottoman Empire* (New York: Cambridge University Press, 2005).

Antweiler, Christoph. *Inclusive Humanism. Anthropological Basics for a Realistic Cosmopolitanism* (Göttingen: V&R Unipress, and Taipei: National Taiwan University Press, 2012).

Aris, Michael. *Lamas, Princes, and Brigands: Joseph Rock's Photographs of the Tibetan Borderlands of China* (New York: Chinese Institute in America, 1992).

Barthorp, Michael. *Marlborough's Army, 1702–11* (London: Osprey, 1980).

Bawden, Charles R. *The Modern History of Mongolia* (London: Weidenfeld and Nicholson, 1968).

Beal, Edwin George Jr. *The Origin of Likin, 1853–1864* (Harvard: Harvard University Press, 1958).

Bishop, J. F. [i.e. Isabella L. Bird]. *The Yangtze Valley and Beyond: An Account of Journeys in China, Chiefly in the Province of Sze Chuan and Among the Man-Tze of the Somo Territory* (London: Murray, 1899; repr. by Taipei: Ch'eng Wen, 1972).

Black, Jeremy. *Rethinking Military History* (New York: Routledge, 2004).

Brewer, John. *The Sinews of Power: War, Money and the English State, 1688–1783* (London: Hyman, 1989).

Brunnert, H[ippolit] S[emenovich] and V. V. Hagelstrom. *Present Day Political Organization of China* (London: Kelly and Walsh, 1912; repr. by Taipei: Ch'eng Wen, 1971; originally published in Russian in 1910).

Burger, Werner. "Coin Production during the Qianlong and Jiaqing Reigns (1736–1820): Issues in Cash and Silver Supply", in *Metals, Monies, and Markets in Early Modern Societies: East Asian and Global Perspectives*, ed. by Thomas Hirzel and Nanny Kim (Berlin: LIT Verlag, 2008), pp. 171–190.

Burrer, Friedrich, and Holger Müller, eds. *Kriegskosten und Kriegsfinanzierung in der Antike.* [The cost of war and war finance in antiquity.] (Darmstadt: Wissenschaftliche Buchgesellschaft, 2008).

Carter, Alice. *The Dutch Republic in Europe in the Seven Years War* (London: Macmillan, 1971).

Chang, Michael G. *A Court on Horseback: Imperial Touring and the Construction of Qing Rule, 1680–1785* (Cambridge: Harvard University Press, 2007).

Chang Te-Ch'ang. "The Economic Role of the Imperial Household in the Ch'ing Dynasty", in *Journal of Asian Studies* 31/2 (1972), pp. 243–273.

Chartrand, René. *Napoleon's Guns 1792–1815 (1): Field Artillery* (Oxford: Osprey, 2003).

——. *Napoleon's Guns 1792–1815 (2): Heavy and Siege Artillery* (Oxford: Osprey, 2003).

Ch'iu A. K'aiming. "Chinese Historical Documents of the Ch'ing Dynasty, 1644–1911", in *Pacific Historical Review* 1, 3 (1932), pp. 324–336.

Ch'ü Tung-tsu. *Local Government in China under the Ch'ing* (Cambridge, Mass.: Harvard University Press, 1962).

Cipolla, Carlo M. *Guns and Sails in the Early Phase of European Expansion, 1400–1700* (London: Collins, 1965). [Also known under the title *Guns, Sails and Empires; Technological Innovation and the Early Phases of European Expansion.*]

di Cosmo, Nicola (2002). *Warfare in Inner Asian History: 500–1800.* Leiden: Brill.

——. "Did Guns Matter? Firearms and the Qing Formation", in *The Qing Formation in World-Historical Time*, ed. by Lynn Struve (Cambridge, Mass.: Harvard University Press, 2004), pp. 121–166.

——. *Diary of a Manchu Soldier in Seventeenth-Century China* (London: RoutledgeCurzon, 2006).

——. *Military Culture in Imperial China* (Cambridge, Mass.: Harvard University Press, 2009).

di Cosmo, Nicola, and Don H. Wyatt, eds. *Political Frontiers, Ethnic Boundaries and Human Geographies in Chinese History.* (London: RoutledgeCurzon, 2003).

van Creveld, Martin. *Supplying War: Logistics from Wallenstein to Patton* (Cambridge: Cambridge University Press, 1977).

——. *Command in War* (Cambridge: Cambridge University Press, 1985).

——. *Technology and War: from 2000 B.C. to the Present* (London: Macmillan, 1989).

Crossley, Pamela Kyle. *Orphan Warriors: Three Manchu Generations and the End of the Qing World* (Princeton: Princeton University Press, 1990).

——. *The Manchus* (Cambridge, Mass.: Blackwell, 1997).

——. *The Translucent Mirror: History and Identity in Qing Imperial Ideology* (Berkeley: University of California Press, 1999).

Crossley, Pamela Kyle, Helen F. Siu, and Donald S. Sutton. *Empire at the Margins: Culture, Ethnicity, and Frontier in Early Modern China* (Berkeley: University of California Press, 2006).

Dai Yingcong. "To Nourish A Strong Military: Kangxi's Preferential Treatment of His Military Officials", in *War and Society* 18, 2 (2000), pp. 71–91 (Repr. in *Warfare in China Since 1600*, ed. by Kenneth Swope [London: Ashgate, 2005]).

——. "The Qing State, Merchants, and the Military Labor Force in the Jinchuan Campaigns", in *Late Imperial China* 22/2 (2001), pp. 35–90.

——. "Yingyun shengxi: Military Entrepreneurship in the High Qing Period, 1700–1800", in *Late Imperial China* 26/2 (2005), pp. 1–67.

——. *The Sichuan Frontier and Tibet: Imperial Strategy in the Early Qing* (Seattle: University of Washington Press, 2009).

——. "Military Finance of the High Qing Period: An Overview", in *Military Culture in Imperial China*, ed. by Nicola di Cosmo (Cambridge, Mass.: Harvard University Press, 2009), pp. 296–316.

——. "Civilians Go into Battle: Hired Militias in the White Lotus War, 1796–1805", in *Asia Maior*, Third Series, 22, II (2009), pp. 145–178.

Dickson, Peter George Muir. *The Financial Revolution in England: A Study in the Development of Public Credit 1688–1756* (London: Macmillan, 1967).

Doyle, William. *Venality: The Sale of Offices in Eighteenth-Century France* (Oxford: Oxford University Press, 1996).

Drelichman, Mauricio, and Hans-Joachim Voth. "Debt Sustainability in Historical Perspective: The Role of Fiscal Repression", in *Journal of the European Economic Association* 6 (2008), pp. 657–667.

Elgood, Robert. *Firearms of the Islamic World in the Tareq Rajab Museum, Kuwait* (London: Tauris, 1995).

Elliott, Mark C. *The Manchu Way: The Eight Banners and Ethnic Identity in Later Imperial China* (Stanford: Stanford University Press, 2001).

Elliott, Mark C., and Peter Stearns. *Emperor Qianlong: Son of Heaven, Man of the World* (Upper Saddle River: Longman, 2009).

Entenmann, Robert. "The Myth of Sonom, the Hmong King", in *Hmong Studies Journal* 6 (2005), pp. 1–14.

Ertman, Thomas. *Birth of the Leviathan: Building States and Regimes in Medieval and Early Modern Europe* (Cambridge: Cambridge University Press, 1997).

Fairbank, John King, and Ssu Yü Teng. "On the Transmission of Ch'ing Documents", in *Harvard Journal of Asiatic Studies* 4 (1939), pp. 12–46.

——. "On the Types and Uses of Ch'ing Documents", in *Harvard Journal of Asiatic Studies* 5 (1940), pp. 1–71.

Fang Chaoying. "A Technique for Estimating the Numerical Strength of the Early Manchu Military Forces", in *Harvard Journal of Asiatic Studies* 13 (1950), pp. 192–215.

Farquhar, David M. "Emperor as Bodhisattva in the Governance of the Ch'ing Empire", in *Harvard Journal of Asiatic Studies* 38 (1978), pp. 5–34.

Gardella, Robert. "Squaring Accounts: Commercial Bookkeeping Methods and Capitalist Rationalism in Late Qing and Republican China", in *Journal of Asian Studies* 51 (1992), pp. 317–339.

Giersch, C. Patterson. *Asian Borderlands: The Transformation of Qing China's Yunnan Frontier* (Cambridge. Mass.: Harvard University Press, 2006).

von Glahn, Richard. *Fountain of Fortune: Money and Monetary Policy in China, 1000–1700* (Berkeley: University of California Press, 1996).

Gordon, Raymond G. *Ethnologue: Languages of the World* (Dallas: SIL International, 2005).

Greatrex, Roger. "A Brief Introduction to the First Jinchuan War (1747–1749)", in *Tibetan Studies: Proceedings of the 6th Seminar of the International Association for Tibetan Studies, Fagernes 1992*, ed. by Per Kværne (Oslo: The Institute for Comparative Research in Human Culture, 1994), pp. 247–263.

Guy, R. Kent. *Qing Governors and Their Provinces: The Evolution of Territorial Administration in China, 1644–1796* (Seattle: University of Washington Press, 2010).

Haenisch, Ernst. "Das Goldstromland im chinesisch-tibetischen Grenzgebiete, nach dem großem Kriegswerk vom Jahre 1781 dargestellt" [The Gold-River country, i.e. Jinchuan, in the China-Tibet border area, according to the war history from 1781], in *Southern Tibet*, Vol. IX, ed. by Sven Hedin and Albert Herrmann (Stockholm: Lithographic Institute of the General Staff [...], 1922), pp. 67–132.

——. "Das *Ts'ing-shih-kao* und die sonstige chinesische Literatur zur Geschichte der letzten 300 Jahre" [The *Qingshigao* and other documentary sources on the Chinese history of the past 300 years], in *Asia Maior* VI (1930), pp. 403–444.

——. "Die Eroberung des Goldstromlandes in Ost-Tibet als Beitrag zur chinesischen Kolonialgeschichte des 18. Jahrhunderts" [The conquest of the Gold-River country, i.e. Jinchuan, in Eastern Tibet: a contribution to Chinese colonial history], in *Asia Maior* X (1935), pp. 17–21.

Harding, David, and the Diagram Group, eds. *Weapons: An International Encyclopedia from 5000 B.C. to 2000 A.D.* (New York: St. Martin's Press, 1980).

Hartill, David. *Qing Cash* (London: Royal Numismatic Society, 2003).

Hsü, Immanuel C. Y. *The Rise of Modern China* (New York/Oxford: Oxford University Press, 2000[6]; first edition 1970).

Hucker, Charles O. *China's Imperial Past: An Introduction to Chinese History and Culture* (Stanford: Stanford University Press, 1975).

——. *A Dictionary of Official Titles in Imperial China* (Stanford: Stanford University Press, 1985).

Hummel, Arthur W. *Eminent Chinese of the Ch'ing Period* (Taipei: SMC Publishing, 1991; first edition Washington: US Government Printing Office, 1944).

Jacques, Guillaume. *Phonologie et morphologie du Japhug (rGyalrong)* [Phonology and morphology of the Japhug (rGyalrong) language] (Ph. Diss.: Paris, 2004).

Jones, D. W. *War and Economy in the Age of William III and Marlborough* (Oxford: Blackwell, 1988).

Jones, Eric L. *The European Miracle: Environments, Economies, and Geopolitics in the History of Europe and Asia* (Cambridge: Cambridge University Press, 1981).

Kaske, Elisabeth. "The Price of an Office: Venality, the Individual and the State in 19th Century China", in *Metals, Monies, and Markets in Early Modern Societies: East Asian and Global Perspectives*, ed. by Thomas Hirzel and Nanny Kim (Berlin: LIT Verlag, 2008), pp. 281–308.

——. "Fund-Raising Wars: Office Selling and Interprovincial Finance in Nineteenth Century China", in *Harvard Journal of Asiatic Studies* 71/1 (2011), pp. 69–141.

Keegan, John. *A History of Warfare* (New York: Knopf, 1993).

Kim, Nanny. "Copper Transports Out of Yunnan, ca. 1750–1850: Transport Technologies, Natural Difficulties and Environmental Change in a Southwestern Highland Area", in *Metals, Monies, and Markets in Early Modern Societies: East Asian and Global Perspectives*, ed. by Thomas Hirzel and Nanny Kim (Berlin: LIT Verlag, 2008), pp. 191–220.

King, Frank H. H. *Money and Monetary Policy in China, 1845–1895* (Cambridge, Mass.: Harvard University Press, 1965).

Kiser, Edgar, and Yong Cai. "War and Bureaucratization in Qin [!] China: Exploring an Anomalous Case", in *American Sociological Review* 68/4 (2003), pp. 511–539.

Knox, MacGregor and Williamson Murray. *The Dynamics of Military Revolution 1300–2050* (Cambridge: Cambridge University Press, 2001).

Kwan, Man Bun. *The Salt Merchants of Tianjin: State-Making and Civil Society in Late Imperial China* (Honolulu: University of Hawai'i Press, 2001).

Lee Kuo-chi. "Bemerkungen zum Ch'ing-Dokumentenstil: Eingaben an den Kaiser (tsou)" [Some remarks on the style in Qing documents: Memorials (*zou*)], in *Oriens Extremus* 17 (1970), pp. 125–135.

Leonard, Jane Kate. "Stretching the Bureaucracy: Ad Hoc Agencies and Recruitment in the 1826 Sea Transport Experiment", in *Metals, Monies, and Markets in Early Modern Societies: East Asian and Global Perspectives*, ed. by Thomas Hirzel and Nanny Kim (Berlin: LIT Verlag, 2008), pp. 255–280.

Leonard, Jane Kate, and John R. Watt, eds. *To Achieve Security and Wealth: The Qing Imperial State and the Economy, 1644–1911* (Ithaca, Cornell University, 1992).

Li, Lillian M. "Grain Prices in Zhili Province, 1736–1911: A Preliminary Study", in *Chinese History in Economic Perspective*, ed. by Thomas G. Rawski and Lillian M. Li (Berkeley: University of California Press, 1992), pp. 69–99.

Liang Jieming [i.e. Leong Kit Meng]. *Chinese Siege Warfare: Mechanical Artillery and Siege Weapons of Antiquity, an Illustrated History* (Singapore: published by the author, 2006).

Lin Man-houng. *China Upside Down: Currency, Society, and Ideologies, 1808–1856* (Cambridge, Mass.: Harvard University Press, 2006).

Liu, Kwang-Ching. "Chinese Merchant Guilds: An Historical Inquiry", in *Pacific Historical Review*, 57/1 (1988), pp. 1–23.

Liu Mau-Tsai. "Struktur und Terminologie chinesischer Schriftstücke im Behördenverkehr" [Structure of and terminology in Chinese administrative documents], in *Nachrichten der Gesellschaft für Natur- und Völkerkunde Ostasiens* [Proceedings of the Society for East Asian Ethnology] 109 (1971), pp. 5–46.

Liu Ts'ui-jung. "An Analysis of the Land Tax Burden in China, 1650–1865", in *Journal of Economic History* 37/2 (1977), pp. 359–381.

von Loehr, A. O. "Die Finanzierung des Siebenjährigen Krieges: Ein Versuch vergleichender Geldgeschichte" [Funding of the Seven Years' War: An attempt on comparative monetary history], in *Numismatische Zeitschrift* [Numismatics magazine] XVIII (1925), pp. 95–110.

Lorge, Peter A. *War, Politics and Society in Early Modern China, 900–1795* (London: Routledge, 2005).

——. *Warfare in China to 1600* (Aldershot: Ashgate, 2005).

——. *The Asian Military Revolution: From Gunpowder to the Bomb* (Cambrige, Mass.: Cambridge University Press, 2008).

Lynn, John A. *Feeding Mars: Logistics in Western Warfare from the Middle Ages to the Present* (Boulder: Westview, 1993).

——. *The Wars of Louis XIV, 1667–1714* (London: Longman, 1999).
——. *Battle: A History of Combat and Culture* (Boulder: Westview, 2004).
Mann, Susan. *Local Merchants and the Bureaucracy, 1750–1950*. Standford: Stanford University Press, 1987.
Mann Jones, Susan. "Finance in Ning-po: The 'Ch'ien Chuang', 1750–1880", in *Economic Organisation in Chinese Society*, ed. by W. E. Willmott (Stanford, Cf.: Stanford University Press, 1972), pp. 47–78.
Mansier, Patrick. "La guerre du Jinchuan (rGyal-rong): Son contexte politico-religieux" [The politico-religious context of the Jinchuan (rGyal-rong) war], in *Tibet: Civilisation et Société, Colloque organisé par la Fondation Singer-Polignac à Paris, les 27, 28, 29 avril 1987* [Civilization and Society of Tibet: A symposium organised by the Singer-Polignac Foundation] (Paris: Éditions de la Maison des sciences de l'homme, 1990), pp. 125–142.
Marichal, Carlos. *Bankruptcy of Empire: Mexican Silver and the Wars Between Spain, Britain and France, 1780–1810* (Cambridge: Cambridge University Press).
Martin, Dan. "Bonpo Canons and Jesuit Cannons: On Sectarian Factors Involved in the Ch'ien-lung Emperor's Second Goldstream Expedition of 1771–1776", in *The Tibet Journal* 15/2 (1990), pp. 3–28.
Mayers, William Frederick. *The Chinese Government: A Manual of Chinese Titles, Categorically Arranged and Explained* (London: Kelly & Walsh, 1897; repr. Taipei: Ch'eng-wen, 1970).
McElderry, Andrea Lee. *Shanghai's Old Style Banks (Ch'ien Chuang), 1800–1935: A Traditional Institution in a Changing Society* (Ann Arbor: Center for Chinese Studies, University of Michigan: 1976).
McNeill, William. *The Pursuit of Power: Technology, Armed Force and Society since A.D. 1000* (Chicago: University of Chicago, 1982).
Metzger, Thomas A. *The Internal Organisation of the Ch'ing Bureaucracy: Legal, Normative, and Communication Aspects* (Cambridge: Harvard University Press, 1973).
Millward, James A. *Beyond the Pass: Economy, Ethnicity, and Empire in Qing Central Asia, 1759–1864* (Stanford: Stanford University Press, 1998).
——. *New Imperial History: The Making of Inner Asian Empire at Qing Chengde* (London: Routledge, 2004).
Murphey, Rhoads. *Ottoman Warfare, 1500–1700* (New Brunswick, N.J.: Rutgers University Press, 1999).
Naquin, Susan. *Shantung Rebellion: The Wang Lun Uprising of 1774* (New Haven: Yale University Press, 1981).
Neal, Larry. "Interpreting Power and Profit in Economic History: A Case Study of the Seven Years War", in *Journal of Economic History* XXXVII/1 (1977), pp. 20–35.
——, ed. *War Finance* (Aldershot: Edward Elgar, 1994).
Needham, Joseph. *Science and Civilization in China*, vol. 5, *Chemistry and Chemical Technology*, part 7, *Military Technology; the Gunpowder Epic* (Cambridge: Cambridge University Press, 1962).
——. *Science and Civilization in China*, vol. 4, *Physics and Physical Technology*, part 1, *Physics* (Cambridge: Cambridge University Press, 1986).
Newby, Laura J. *The Empire and the Khanate: A Political History of Qing Relations With Kokand, c. 1760–1860* (Leiden: Brill, 2005).
Olbricht, Peter. *Das Postwesen in China unter der Mongolenherrschaft im 13. und 14. Jahrhundert* [The postal system in China under the Mongol rule, 13th-14th centuries] (Wiesbaden: Harrassowitz, 1954).
O'Sullivan, Patrick, and Jesse W. Miller Jr. *The Geography of Warfare* (London, Canberra: Croom Helm, 1983).
Park, Nancy E. "Corruption in Eighteenth-Century China", in *Journal of Asian Studies* 56/4 (1997), pp. 967–1005.
Parker, Geoffrey. *The Military Revolution: Military Innovation and the Rise of the West, 1500–1800* (Cambridge: Cambridge University Press, 1988).

Peers, Chris. *Late Imperial Chinese Armies, 1520–1840* (London: Osprey, 1997).

Perdue, Peter C. "The Qing State and the Gansu Grain Market, 1739–1864", in *Chinese History in Economic Perspective*, ed. by Thomas G. Rawski and Lillian M. Li (Berkeley: University of California Press, 1992), pp. 100–125.

——. "Military Mobilization in Seventeenth and Eighteenth-Century China, Russia, and Mongolia", in *Modern Asian Studies* 30/4 (1996), *Special Issue: War in Modern China*, pp. 757–793.

——. *China Marches West: The Qing Conquest of Central Eurasia* (Cambridge: Harvard University Press, 2005).

Pines, Yuri. *The Everlasting Empire: The Political Culture of Ancient China and Its Imperial Legacy* (Princeton/Oxford: Princeton University Press, 2012).

Pomeranz, Kenneth. *The Great Divergence: China, Europe, and the Making of the Modern World Economy* (Princeton: Princeton University Press, 2000).

Redlich, Fritz. *The German Military Enterpriser* [sic] *and his Work Force: A Study in European Economic and Social History* (Wiesbaden: Steiner, 1964). [Esp. pp. 239–270.]

Riley, James C. *The Seven Years War and the Old Regime in France: The Economic and Financial Toll* (Princeton: Princeton University Press, 1986).

Rockhill, William Woodville. *The Land of the Lamas: Notes of a Journey Through China, Mongolia and Tibet, with Maps and Illustrations* (New Delhi: Asian Publication Services, 1975; first published by New York: The Century, 1891).

Rockhill, W. W. "The Dalai Lamas of Lhasa and their Relations with the Manchu Emperors of China, 1644–1908", in *T'oung Pao* XI (1910), pp. 1–104 (republ. by Leyden: Brill, 1910; Dharamsala: Library of Tibetan Works and Archives, 1998).

Rogers, Clifford J. *The Military Revolution Debate: Readings on the Military Transformation of Early Modern Europe* (Boulder: Westview, 1995).

Roseveare, Henry. *The Financial Revolution 1660–1760* (London: Longman, 1991).

Roth, Jonathan P. *The Logistics of the Roman Army at War, 264 BC–AD 235* (Leiden: Brill, 2012).

Rowlands, Guy. *The Financial Decline of a Great Power: War, Influence, and Money in Louis XIV's France* (Oxford: Oxford University Press, 2012).

Sawyer, Ralph D. *The Seven Military Classics of Ancient China* (Boulder: Westview, 1993).

Smith, Paul Jakov. "Shen-tsung's Reign and the New Policies of Wang An-shih", in *The Cambridge History of China*, vol. 5, part 1, *The Sung Dynasty and Its Precursors, 907–1279*, ed. by Denis Twitchett and Paul Jakov Smith (Cambridge: Cambridge University Press, 2009), pp. 347–483.

Stötzner, Walther. *Ins unerforschte Tibet: Tagebuch der deutschen Expedition Stötzner 1914* [Into the uncharted land of Tibet: The diary of the German Stötzner expedition in 1914] (Leipzig: Koehler, 1924).

Sun, E-tu Zen. "The Board of Revenue in Nineteenth-Century China", in *Harvard Journal of Asiatic Studies* 24 (1963), pp. 175–228.

Swope, Kenneth, ed. *Warfare in China Since 1600* (Aldershot: Ashgate, 2005).

Tafel, Albert. *Meine Tibetreise: eine Studienfahrt durch das nordwestliche China und die innere Mongolei in das östliche Tibet* [My study trip through Tibet, crossing northwestern China, Inner Mongolia, and Eastern Tibet] (Stuttgart: Union Deutsche Verlagsgesellschaft, 1923).

Teichmann, Eric. *Travels of a Consular Officer in Eastern Tibet: Together with a History of the Relations between China, Tibet and India* (Cambridge: Cambridge University Press, 1922).

Temple, Richard, and Thomas Wade, "Political Lessons of Chinese History", in *Transactions of the Royal Historical Society*, New Series, vol. 1 (1883–1884), pp. 204–241.

Torbert, Preston M. *The Ch'ing Imperial Household Department: A Study of its Organization and Principal Functions, 1662–1796* (Cambridge: Harvard University Press, 1977).

Tsuya, Noriko O. et al., eds. *Prudence and Pressure: Reproduction and Human Agency in Europe and Asia, 1700–1900* (Cambridge, Mass.: MIT, 2010).

Veit, Veronika. "Qalqa 1691 bis 1911" [The Qalqa Mongols between 1691 and 1911], in *Die Mongolen: Beiträge zu ihrer Geschichte und Kultur* [The Mongols: Contributions to their history and culture], ed. by Michael Weiers (Darmstadt: Wissenschaftliche Buchgesellschaft, 1986), pp. 435–466.

Vogel, Hans Ulrich. *Chinese Central Monetary Policy and the Yunnan Copper Mining Industry in the Early Qing, 1644–1800* (Ph. Diss.: Zürich, 1983).

——. "Chinese Central Monetary Policy, 1644–1800", in *Late Imperial China* 8/2 (1987), pp. 1–52.

Vogelsang, Kai. *Geschichte Chinas* [A history of China] (Stuttgart: Reclam, 2012).

Wakeman, Frederic Jr. *The Great Enterprise: The Manchu Reconstruction of Imperial Order in Seventeenth-Century China* (Berkeley: University of California Press, 1985).

Wang Xiangyun. "The Qing Court's Tibet Connection: Lcang skya Rol pa'i rdo rje and the Qianlong Emperor", in *Harvard Journal of Asian Studies* 60/1 (2000), pp. 125–163.

Waley-Cohen, Joanna. "God and Guns in Eighteenth-Century China: Jesuit Missionaries and the Military Campaigns of the Qianlong Emperor (1736–1795)", in *Contact between Cultures*, vol. 4, *Eastern Asia: History and Social Sciences*, ed. by Bernard H. K. Luk (Lewiston: Mellen, 1992), pp. 94–99.

——. "China and Western Technology in the Late Eighteenth Century", in *American Historical Review* 99/4 (1993), pp. 1525–1544.

——. *The Culture of War in China: Empire and the Military under the Qing Dynasty* (London and New York: Tauris, 2006).

——. "Militarization of Culture in Eighteenth-Century China", in *Military Culture in Imperial China*, ed. by Nicola di Cosmo (Cambridge: Harvard University Press, 2009), pp. 278–295. (First published as "On the Militarization of Culture in the Eighteenth-Century Qing Empire", in *Common Knowledge* 12 [2006], pp. 96–106.)

Wang Yeh-chien. *An Estimate of the Land-Tax Collection in China, 1753 and 1908* (Cambridge: Harvard University Press, 1973).

——. *Land Taxation in Imperial China, 1750–1911* (Cambridge: Harvard University Press, 1973).

Wennerlind, Carl. *Casualties of Credit: The English Financial Revolution, 1620–1720* (Cambridge, Mass.: Harvard University Press, 2011).

Wilkinson, Endymion. *Chinese History: A Manual* (Cambridge: Harvard University Press, 2000).

Will, Pierre Étienne. *Bureaucratie et famine en Chine au 18e siècle* (Paris: Mouton, 1980; also published as *Bureaucracy and famine in eighteenth-century China*, Stanford: Stanford University Press, 1990).

Will, Pierre Étienne, and R. Bin Wong. *Nourish the People: The State Civilian Granary System in China, 1650–1850* (Ann Arbor: University of Michigan Press, 1991).

Woodside, Alexander. "The Ch'ien-lung Reign", in *Cambridge History of China*, vol. 9, *The Ch'ing Empire to 1800*, part 1, ed. by Willard J. Peterson (Cambridge: Cambridge University Press, 2002), pp. 230–309.

Yang Lien-sheng. *Money and Credit in China: A Short History* (Cambridge, Mass.: Harvard University Press, 1952).

Zelin, Madeleine. *The Magistrate's Tael: Rationalizing Fiscal Reform in Eighteenth-Century China* (Berkeley: University of California Press, 1984).

Zhang Hongxiang. "Studies in Late Qing Dynasty Battle Paintings", in *Artibus Asiae* 60/2 (2000), pp. 265–296.

Zheng Qin. "Pursuing Perfection: Formation of the Qing Code", in *Modern China* 21 (1995), pp. 310–344.

Authors and Editors in Chinese

Cai Renzheng 蔡仁政. "Jinchuan Zangzu shibing baowei zuguo de zhandou yeji jianshu" 金川藏族士兵保衛祖國的戰鬪業績簡述 [A short account of the glorious fight of

Tibetan soldiers from Jinchuan defending the motherland], in *Xizang yanjiu* 西藏研究 1993/2, pp. 75–79.

Cao Qifu 曹啓富. "Lüelun Qianlong nianjian Da-Xiao Jinchuan zhi yi" 略論乾隆年間大小金川之役 [A short discussion of the war against Greater and Smaller Jinchuan during the Qianlong reign], in *Sichuan shifan xueyuan xuebao* (*zhexue shehui kexue ban*) 四川師範學院學報 (哲學社會科學版) 1999/6, pp. 39–42.

Chen Feng 陳鋒. *Qingdai junfei yanjiu* 清代軍費研究 [Military expenditure in Qing China] (Wuchang: Wuhan daxue chubanshe, 1992).

——. "Qingdai caizheng zhichu zhengce yu zhichu jiegou de biandong" 清代財政支出政策與支出結構的變動 [The politics of government expenditure and its structural change during the Qing period], in *Jiang-Han luntan* 江漢論壇 2000/5, 60–70.

Chen Hua 陳華. "*Sizhu qingce* yu xiandai kuaiji"《四柱清冊》與現代會計 [The four columns clearance book method and modern accounting], in *Zhongguo nongye kuaiji* 中國農業會計 1997/4, pp. 47–48.

Chen Li 陳力. "*Jinchuan jilüe* ji qi xiangguan wenti"《金川紀略》及其相關問題 [About some problems concerning the book *Jinchuan jilüe*], in *Sichuan daxue xuebao* (*zhexue shehui kexue ban*) 四川大學學寶 (哲學社會科學版) 1992/3, pp. 90–94.

Chen Shisong 陳世松, ed. *Sichuan tongshi* 四川通史 [A comprehensive history of Sichuan] (Chengdu: Sichuan daxue chubanshe, 1992).

Chen Xiaoqiang 陳小強. "Jinchuan zhi luan hou Qingjun zhong de Zangzu jiangling chutan" 金川之亂後清軍中的藏族將領初探 [A first research about Tibetan generals and officers in the Qing army after the end of the Jinchuan rebellion], in: *Zhongguo zangxue* (*Hanwen ban*) 中國藏學 (漢文版) 1994/4, pp. 105–113.

Cheng Dong 成東. "Qingdai huoqi" 清代火器 [Firearms of the Qing period], in *Zhongguo da baike quanshu: Wenwu, bowuguan* 中國大百科全書：文物博物館 [Encyclopedia Sinica, Artefacts and museums volume] (Beijing/Shanghai: Zhongguo da baike quanshu chubanshe, 1993), pp. 426–428.

Dai Yi 戴逸, and Hua Li 華立. "Yi chang de bu chang shi de zhanzheng: Lun Qianlong chao Jinchuan zhi yi" 一場得不償失的戰爭：論乾隆朝金川之役 [A war that did not pay off: A discussion of the Jinchuan wars during the Qianlong reign], in *Lishi yanjiu* 歷史研究 1993/3, pp. 30–41.

Dai Yi 戴逸, and Zhang Shiming 張世明. *Shiba shiji de Zhongguo yu shijie: Junshi juan* 十八世紀的中國與世界：軍事卷 [China and the world in the 18th century: Military volume] (Liaoyang: Liaohai chubanshe, 1998).

Deng Shaohui鄧紹輝. "Xian-Tong shiqi zhongyang yu difang caizheng guanxi de yanbian" 咸同時期中央與地方財政關係的演變 [The development of the financial politics between central and local governments during the Xianfeng and Tongzhi reigns], in *Shixue yuekan* 史學月刊 2001/3, pp. 52–59.

Fang Baozhang 方寶璋. "Zhongguo gudai shenji shi gailun" 中國古代審計史概論 [An outline of the history of accounting in ancient China], in *Zhongguo shi yanjiu* 中國史研究 1996/1, pp. 3–17.

Fu Zhongxia 傅仲俠 et al., eds. *Zhongguo junshi shi* 中國軍事史 [Military history of China], *Fujuan: Lidai zhanzheng nianbiao* 附卷：歷代戰爭年表 [Appendix volume: Chronological table of wars through history] (Beijing: Jiefangjun chubanshe, 1986), vol. 2.

He Xuefei 何學飛. "Jianxi kuaijifa de yanjin ji qi shishi" 簡析會計法的演進及其實施 [A short analysis of the development of accounting and its practice], in *Hunan caijing gaodeng zhuanke xuexue xiaobao* 湖南財經高等專科學校學報 2003/6, pp. 75–78.

Huang Da 黃達, Liu Hongru 劉鴻儒, and Zhang Xiao 張肖, eds. *Zhongguo jinrong baike quanshu* 中國金融百科全書 [Encyclopedia of money and finance in China] (Beijing: Jingji guanli chubanshe, 1990).

Huang Huixian 黃惠賢, and Chen Feng 陳鋒. *Zhongguo fenglu zhidu shi* 中國俸祿制度史 [History of the state salary system in China] (Wuhan: Wuhan daxue chubanshe, 2005).

Huang Tianhua 黃天華. *Zhongguo caizheng shigang* 中國財政史綱 [An outline of financial history in China] (Shanghai: Shanghai caijing daxue chubanshe, 1999).

Jiang Guozhu 姜國柱, ed. *Zhongguo junshi sixiang tongshi* 中國軍事思想通史 [A comprehensive history of military thought in China] (Beijing: Zhongguo shehui kexue chubanshe, 2006).

Kōsaka Masanori 香坂昌紀. "Qingdai qianqi dui Dzunghar zuozhan de jingji xiaoguo" 清代前期對准噶爾作戰的經濟效果 [The economic repercussions of the wars against the Dzunghars in the early Qing period], transl. by Li Xiaolin 李小林, in *Shixue jikan* 史學集刊 2000/4, pp. 17–24.

Lai Fushun 賴福順. *Qianlong zhongyao zhanzheng zhi junxu yanjiu* 乾隆重要戰爭之軍需研究 [The expenditure for the great wars of the Qianlong emperor] (Taibei: Gugong bowuyuan, 1984).

Lan Yong 藍勇. *Xinan lishi wenhua dili* 西南歷史文化地理 [Historical and cultural geography of the southwest] (Chongqing: Xinan shifan daxue chubanshe, 1997).

Lei Bingyan 雷炳炎. "Qingchu jungong fengjue zhidu shulun" 清初軍功封爵制度述論 [An essay on the system of nobility titles for military merits in the early Qing period], in *Hu-Xiang luntan* 湖湘論壇, 127/4 (2009), pp. 121–124

Lei Qing 雷慶, ed. *Zhongguo bingxue wenhua mingzhu* 中國兵學文化名著 [Famous writings of Chinese military culture] (Yanji: Yanbian daxue chubanshe, 1995).

Li Hongbin 李鴻彬, and Bai Jie 白傑. "Ping Qianlong chao Jinchuan zhi yi" 評乾隆朝金川之役 [A critique of the Jinchuan wars during the Qianlong reign], in *Qingshi yanjiu* 清史研究 1998/2, pp. 66–76.

Li Wei 李巍. "*Fuheng zongci bei* yu Da Jinchuan zhanyi"《傅恆宗祠碑》與大金川戰役 [The 'Stele of the ancestral altar of Fuheng' and the war against Greater Jinchuan], in *Beijing wenbo* 北京文博 1999/1, pp. 84–88.

Li Zhongqing 李中清. *Zhongguo xinan bianjiang de shehui jingji: 1250–1850* 中國西南邊疆的社會經濟：1250–1850 [Society and economy in the southwestern borderlands of China, 1250–1850] (Beijing: Renmin chubanshe, 2012).

Liang Yanbing 梁嚴冰, and Liu Rong 劉蓉. "Qingdai de juanna zhidu" 清代的捐納制度 [The contribution system during the Qing period], in *Lishi jiaoxue* 歷史教學 1999/9, pp. 50–53.

Liu Fengyun 劉鳳雲. "Qing Kangxi chao juanna dui lizhi de yingxiang" 清康熙朝捐納對吏治的影響 [The influence of the contributions on the local administration during the Kangxi reign], in *Henan daxue xuebao (shehui kexue ban)* 河南大學學報 (社會科學版) 43 (2003), pp. 6–11.

Lu Xixing 陸錫興, ed. *Zhongguo gudai qiwu da cidian: Bingqi, xingju* 中國古代器物大詞典：兵器, 刑具 [Great dictionary of tools in ancient China: Weapons and penal tools volume] (Shijiazhuang: Henan jiaoyu chubanshe, 2002).

Lu Zijian 魯子健. "Juanna: Qingdai de maiguan yujue zhidu" 捐納—清代的賣官鬻爵制度 [Contributions: The purchase of offices during the Qing period], in *Wenshi zazhi* 文史雜志 1999/6, pp. 74–76.

Luo Ergang 羅爾綱. *Luyingbing zhi* 綠營兵志 [Treatise on the Green Standard system] (Beijing: Zhonghua shuju, 1984).

Luo Yudong. 羅玉東. *Zhongguo lijin shi* 中國釐金史 [History of the *likin* tax] (Shanghai: Shangwu yinshuguan, 1936; repr. in *Jindai Zhongguo shiliao congkan xubian* 近代中國史料叢刊續編, series 62, no. 612, vols. 6–7).

Nie Chongzheng 聶崇正. "Qing pingding Liang Jinchuan gongchen xiang gouzhi" 清平定兩金川功臣像鉤治 [A research into the portraits of deserving Qing officials of the war against the Two Jinchuans], in *Shoucangjia* 收藏家 1995/12, pp. 48–50.

Pan Honggang 潘洪鋼. "Qianlong chao Liang Jinchuan gaitu guiliu zhi xingqi" 乾隆朝兩金川改土歸屯之興起 [The origin of the replacement of autonomous administration by military colonies in the Two Jinchuans during the Qianlong reign], in *Zhongnan minzu xuexuan xuebao* 中南民族學院學報 1988/5, pp. 35–41.

——. "Qingdai Qianlong chao Liang Jinchuan gaitu guitun kao" 清代乾隆朝兩金川改土歸屯考 [An investigation into the replacement of autonomous administration by military colonies in the Two Jinchuans during the Qianlong reign], in *Minzu yanjiu* 民族研究 1988/6, pp. 62–71.

——. "Qingdai guanxiu minzu tuce *Huang-Qing zhigong tu* qianshuo" 清代官修民族圖冊《皇清職貢圖》淺說 [A short explanation of the government-issued illustrated ethnology handbook *Huang-Qing zhigong tu*], in *Zhongnan minzu xueyuan xuebao* (*zhexue shehui kexue ban*) 中南民族學院學報 (哲學社會科學版) 1992/4, pp. 87–90.

Peng Jianying 彭建英. *Zhongguo gudai jimi zhengce de yanbian* 中國古代羈縻政策的演變 [The development of the politics of indirect rule in ancient China] (Beijing: Zhongguo shehui kexue chubanshe, 2004).

Peng Xinwei 彭信威. *Zhongguo huobi shi* 中國貨幣史 [A monetary history of China] (Shanghai: Shanghai renmin chubanshe: 2007; first publ. by Shanghai: Qunlian chubanshe, 1954).

Peng Yushu 彭玉書. "Zhongguo de *sizhu qingce* shi dui shijie kuaiji de dute gongxian" 中國的《四柱清冊》是對世界會計的獨特貢獻 [The particular contribution of China's four column ledger method to the accounting systems worldwide], in *Jiaotong caihui* 交通財會 1995/3, p. 56.

Peng Zhiyan 彭陟焱. "Qianlong chu ding Jinchuan zhanzheng goushen" 乾隆初定金川戰爭鉤沉 [A research into the Qianlong emperor's first war against Jinchuan], in *Xizang minzu xueyuan xuebao* (*zhexue shehui kexue ban*) 西藏民族學院學報 (哲學社會科學版) 2003/7, pp. 94–103.

——. "Shilun Qianlong pingding Jinchuan zhi yingxiang" 試論乾隆平定金川之影響 [On the consequences of the Qianlong emperor's Jinchuan wars], in *Xizang yanjiu* 西藏研究 2003/1, pp. 82–86.

——. "Qianlong zai ding Liang Jinchuan zhanzheng goushen" 乾隆再定兩金川戰爭鉤沈 [Research into the second war against the Two Jinchuans during the Qianlong reign], in *Xizang minzu xueyuan xuebao* (*zhexue shehui kexue ban*) 西藏民族學院學報 (哲學社會科學版) 2004/3, pp. 22–28.

——. "Xi Qianlong di zai Da-Xiao Jinchuan diqu tuixing de tuntian zhidu" 析乾隆帝在大小金川地區推行的屯田制度 [An analysis of the Qianlong emperor's application of the military colony system in the region of Greater and Lesser Jinchuan], in *Xizang minzu xueyuan xuebao* (*zhexue shehui kexue ban*) 西藏民族學院學報 (哲學社會科學版) 2006/9, pp. 26–28, 41.

——. "Qianlong di dui Da-Xiao Jinchuan tusi gaitu guiliu xi" 乾隆帝對大小金川土司改土歸流析 [An analysis of the Qianlong emperor's conversion of autonomous administration into military colonies in Greater and Lesser Jinchuan], in *Xizang minzu xueyuan xuebao* (*zhexue shehui kexue ban*) 西藏民族學院學報 (哲學社會科學版) 2007/7, pp. 15–18, 23.

Qiu Guangming 丘光明, Qiu Long 邱隆, and Yang Ping 楊平. *Zhongguo kexue jishu shi: Duliangheng juan* 中國科學技術史：度量衡卷 [History of Chinese science and technology: Weights and measures volume] (Beijing: Kexue chubanshe, 2001).

Ren Zhiyong 任智勇. "Shishu wan-Qing hubu yinku zhidu yu gengzi zhi hou de biange" 試述晚清戶部銀庫制度與庚子之後的變革 [A tentative description of the treasury of the Ministry of Revenue in late Qing China and the Boxer uprising in the year *gengzi*, i.e. 1900], in *Qingshi yanjiu* 清史研究 2005/8, pp. 44–50.

Rui Helin 芮和. "Qianxi Qianlong shiqi Changlu yanshang zouxiang shuailuo de yuanyin" 淺析乾隆時期長蘆鹽商走向衰落的原因 [A short analysis of the reasons of the decline of the Changlu salt merchants during the Qianlong reign], in *Yanyeshi yanjiu* 鹽業史研究 1994/4, pp. 22–26.

Shen Xuefeng 申學鋒. "Qingdai caizheng zhichu guimo yu jiegou yanbian shulüe" 清代財政支出規模與結構演變述略 [A short account of the dimensions of government expenditure and its structural change during the Qing period], in *Xueshu yanjiu* 學術研究 2004/7, pp. 98–102.

Tan Qixiang 譚其驤, ed. *Zhongguo lishi ditu ji* 中國歷史地圖集 [The historical atlas of China], vol. 8, *Qing shiqi* 清時期 [The Qing dynasty period] (Beijing: Zhongguo ditu chubanshe, 1996).

Tong Zhenglun 童正倫, and Zhang Qun 張群. "Tongban *Pingding Liang Jinchuan desheng tu*" 銅版《平定两金川得勝圖》[The bronze plates "Images from the victory in the war against the Two Jinchuans"], in *Zhongguo dianji yu wenhua* 中国典籍与文化 2003/2, pp. 17–18, 20.

Wang Weihai 王威海. *Zhongguo huji zhidu: lishu yu zhengzhi de fenxi* 中國戶籍制度：歷史與政治的分析 [The household system of China: Historical development and policy] (Shanghai: Shanghai wenhua chubanshe, 2005).

Wang Zhaochun 王兆春. *Zhongguo kexue jishu shi: Junshi jishu juan* 中國科學技術史：軍事技術卷 [History of Chinese science and technology: Military technology volume] (Beijing: Kexue chubanshe, 1998).

Xiao Yishan 蕭一山. *Qingdai tongshi* 清代通史 [A comprehensive history of the Qing] (Taibei: Taiwan shangwu yinshuguan, 1963).

Xin Xingbao 邢新寶. "*Lüying* de duyin 綠營的讀音" [The pronunciation of the word 綠營], in *Zhongxue lishi jiaoxue yanjiu* 中學歷史教學研究 2007/1, 2, p. 88 (Also published in *Zhongxue yuwen jiaoxue* 中學語文教學 2007/5, p. 65).

Xu Daling 許大齡. *Qingdai juanna zhidu* 清代捐納制度 [The contribution system during the Qing period] (Xianggang: Longmen shudian, 1968).

Xu Huaibao 徐懷寶. "Qingdai Jinchuan gai tu wei tun" 清代金川改土爲屯 [The replacement of autonomous administration by military colonies in Jinchuan during the Qing period], in *Shoudu shifan daxue xuebao* (*shehui kexue ban*) 首都師範大學學報 (社會科學版) 1995/5, pp. 56–63.

Xu Ming 徐銘. "Benjiao yu Da-Xiao Jinchuan zhanzheng" 笨教與大小金川戰爭 [The Bön school and the war against Greater and Lesser Jinchuan], in *Kangding xuekan* 康定學刊 1997/3, pp. 14–16.

Yang Dongliang 楊東梁, and Zhang Hao 張浩. *Zhongguo Qingdai junshi shi* 中國清代軍事史 [History of Chinese military during the Qing period] (Beijing: Renmin chubanshe, 1994; series *Baijuanben Zhongguo quanshi* 百卷本中國全史 [Chinese history in one hundred volumes]).

Yang Zukai 楊祖愷. "Qingdai guanli juanna zhidu qianxi yu sikao" 清代官吏捐納制度淺析與思考 [Some thoughts about the purchase of offices during the Qing period], in *Guizhou wenshi congkan* 貴州文史叢刊 1999/4, pp. 37–40.

Yin Xiaodong 尹曉冬. "17 shiji chuan Hua Xiyang tongpao zhizao jishu yanjiu: yi *Xifa shenji, Huogong qieyao* wei zhongxin" 17 世紀傳華西銅炮製造技術研究：以《西法神機》、《火攻挈要》爲中心 [A research on the passing on of Western bronze cannons to China during the 17th century, as reflected in the books *Xifan shenji* and *Huogong qieyao*], in *Zhongguo kexue shi zazhi* 中國科學史雜志 30 (2009), 4, pp. 377–395.

Zeng Weiyi 曾唯一. "Qianlong pingding Jinchuan hou de shanhou shiyi" 乾隆平定金川後的善後事宜 [The regulations for the reconstruction after the Jinchuan war during the Qianlong reign], in *Sichuan shifan daxue xuebao* (*shehui kexue ban*) 四川師範大學學報 (社會科學版) 1986/6, pp. 68–74.

Zhang Changfu 張昌富. "Qianlong pingding Jinchuan dui Jiarong wenhua de yingxiang" 乾隆平定金川對嘉絨的影響 [The influence of the Jinchuan wars during the Qianlong reign on the rGyal-rong culture], in *Xizang yishu yanjiu* 西藏藝術研究 1995/2, pp. 72–76.

Zhang Ting 張婷. "Shixi diyici Jinchuan zhanzheng baofa de zhijie yuanyin" 試析第一次金川戰爭爆發的直接原因 [A tentative analysis of the causes for the first Jinchuan war], in *Sichuan daxue xuebao* (*zhexue shehui kexue ban*) 四川大學學報（哲學社會科學版) 2004/1; pp. 48–50.

Zhang Xiaozhong 張孝忠, and Mao Wanliang 毛萬良. "Jinchuan faxian Qing Qianlong nianjian tusi yinzhang" 金川發現清乾隆年間土司印章 [The seal of a native *tusi* ruler from the Qianlong reign discovered in the Jinchuan region], in *Sichuan wenwu* 四川文物 1987/4, p. 38.

Zhang Yuxin 張羽新. "Qingdai qianqi ge minzu tongyi guannian de lishi tezheng" 清代前期各民族統一觀念的歷史特徵 [The historical distinctiveness of the concept of unification of all ethnic groups during the early Qing], in *Qingshi yanjiu* 清史研究 1996/2, pp. 30–38.

Zhou Wei 周維. *Zhongguo bingqi shigao* 中國兵器史稿 [Outline to a history of weapons in China] (Tianjin: Baihua wenyi chubanshe, 2006).

Zhou Zhichu 周志初. *Wan-Qing caizheng jingji* 晚清財政經濟 [Financial politics and the economy in the late Qing] (Jinan: Qi-Lu shushe, 2002).

Zhuang Jifa 莊吉發. "Qing Gaozong liang ding Jinchuan shimo" 清高宗兩定金川始末 [An outline of Emperor Gaozong's two wars against Jinchuan], in *Dalu zazhi* 大陸雜誌 46 (1973)/1, pp. 1–30.

——. *Qingdai zouzhe zhidu* 清代奏摺制度 [The memorial system of the Qing period] (Taibei: Guoli gugong bowuyuan, 1979).

——. *Qing Gaozong shiquan wugong yanjiu* 清高宗十全武功研究 [A research into the ten successful campaigns of Emperor Qing Gaozong] (Taibei: Guoli gugong bowuguan, 1982, repr. by Beijing: Zhonghua shuju 1987).

Corporate Authors in Chinese

Aba Zangzu zizhizhou diming shouce 阿壩藏族自治州地名手冊 [A handbook on place names in the Tibetan Autonomous county of Ngawa], ed. by Aba zhou diming bangongshi 阿壩州地名辦公室 (Chengdu: Xinan minzu xueyuan, 1987).

Ba-Shu wenhua yu xibu Sichuan kaifa 巴蜀文化與西部四川開發 [The culture of Ba and Shu and the opening of Sichuan in the west], ed. by Sichuan sheng renmin zhengfu canshishi 四川省人民政府參事室 and Sichuan sheng wenshi yanjiuguan 四川省文史研究館 (Chengdu: Sichuan renmin chubanshe, 2001).

Hanyu da cidian 漢語大辭典 [Great dictionary of the Chinese language], ed. by Luo Zhufeng 羅竹風 (Shanghai: Cishu chubanshe, 1986–1993).

Ming-Qing dang'an lunwen xuanbian 明清檔案論文選編 [Selected essays on Ming and Qing archives], ed. by Zhongguo diyi lishi dang'anguan 中國第一歷史檔案館 (Beijing: Zhongguo diyi lishi dang'anguan, 1985).

Shuihudi Qinmu zhujian 睡虎地秦墓竹簡 [The bamboo slips of the Qin tomb in Shuihudi], ed. by Shuihudi Qinmu zhujian zhengli xiaozu 睡虎地秦墓竹簡整理小組 (Beijing: Wenwu chubanshe, 1978[1], 1990[2]).

Sichuan wenwu zhi 四川文物志 [Cultural relics of Sichuan], ed. by Sichuan wenwu guanliju 四川文物管理局 (Chengdu: Ba-Shu shushe, 2005).

Sichuan sheng ditu ce 四川省地圖冊 [Pocket size maps ot the province of Sichuan], ed. by Zhongguo ditu chubanshe 中國地圖出版社 (Beijing: Zhongguo ditu chubanshe, 2001).

Sichuan sheng ditu ji 四川省地圖集 [Maps of the province of Sichuan], ed. by Sichuan sheng cehuiju 四川省測會局 (Chengdu: Sichuan sheng cehuiju, 1981).

Zhongguo lidai zhanzheng shi 中國歷代戰爭史 [A history of war in China through the ages], ed. by Sanjun daxue 三軍大學 (Taibei: Liming, 1963).

Zhongguo junshi shi 中國軍事史 [A history of military affairs in China], ed. by Jiefangjun chubanshe 解放軍出版社 (Beijing: Jiefangjun chubanshe, 1986).

LIST OF CHINESE CHARACTERS

Note: This list only includes terms. The characters of personal names, place names and titles of books are to be found in the index.

an 案
anchashi 按察使
anfushi 安撫使
anjia yin 安家銀
bailiang 白糧
Bailianjiao 白蓮教
bangtie 幫帖
bangtie yin 幫帖銀
banli chaiwu zongju 辦理差務總局
banshi dachen 辦事大臣
baoshaanju 寶陝局
baoxiao 報銷
baqi 八旗
batulu 巴圖魯 (Manchu *baturu*)
bazong 把總
beifu 背夫
beilu 北路
bense 本色
bingchai 兵差
bingding 兵丁
bitieshi 筆帖式 (Manchu *bithesi*)
bobo 餑餑
bokuan 撥款
buku 部庫
buyuan 部院
buzhengshi 布政使
caimai ma 採買馬
caiyunju 採運局
canjiang 參將
canling 參領
canzan 參贊
canzan dachen 參贊大臣
caoliang 漕糧
chaiyi 差役
changfu 長夫
changjiancang 常監倉
changmao 長矛
changpingcang 常平倉
changqiang 長槍
changyun 長運
chaomian 炒麵
chongtian pao 衝天炮 or 沖天炮
chu'an 初案
congyi 從役
cun yin 存銀
da pao 大炮
dajiang pao 大將炮
dalisi 大理寺
dan 石
dangshui 當稅
dao yushi 道御史
daofu 道府
daofu dayuan 道府大員
daoku 道庫
daoyuan 道員
dazhan 大站
dianshi 典史
dianshui 典稅
diao 碉
diding 地丁
difangzhi 地方誌 or 地方志
ding'e 定額
dong san sheng 東三省
dou 斗
duchayuan 督察院
duliangdao 督糧道
dusi 都司
ewai lianxiang 額外練餉
fanfu 藩夫
fanglüe 方略
fangyu 防禦
fanku 藩庫
fansi 藩司
fen 分
fengxiang 俸餉
fenpei 分賠
fenpei zhi li 分賠之例
fu canling 副參領
fudutong 副都統
fujiang 副將
fuxiao 附銷
gaitu guiliu 改土歸流
gangshang 綱商
genyi 跟役
gongbu 工部
Gongcheng zeli 工程則例
gongjiang 工匠
gongpai 功牌
gongshi 公事
guan 貫

guanglusi 光祿寺
guanhuan diding 官宦地丁
guanshang 官商
guanshui 關稅
guanyi 官役
guanyun 官運
guanzhai 官寨
guichou 癸丑
guikuan 歸款
gujiang 雇匠
gunyun 滚運
guozijian 國子監
Hanbing 漢兵
Hanjun 漢軍
hanlinyuan 翰林院
heiqian 黑鉛
hong shutong 紅熟銅
honglusi 鴻臚寺
hongqian 紅錢
houbu 候補
hu 斛
huanban 換班
huang shutong 黃熟銅
huangtie 荒鐵
hubu 戶部
huikong 回空
huimian 灰麵
huodan 火彈
huojian 火箭
huoqiying 火器營
huosheng pan 火繩盤
huoxiao 火硝
jiahao 加耗
jiangjun 將軍
jianjiang 箭匠
jianruiying 健銳營
jianzhan 尖站
jiaoxiang 剿餉
jiaxiang yin 甲餉銀
jimi 羈縻
jin 斤
jingfeiju 經費局
jinghuang 淨磺
jinglüe dachen 經略大臣
jingqian 淨鉛
jingsheng 京升
jingxiao 淨硝
jingyuan 京員
jintie 津貼
jishizhong 給事中 [sic]
jiu'an 舊案
jiuguan 舊管
juan 捐
juan feng 捐俸
juanna 捐納
juannaju 捐納局
juanshu 捐輸
juanyin 捐銀
juanzhen 蠲賑
junjichu 軍機處
junxuju 軍需局
kai juan 開捐
kaichu 開除
kaixuan 凱旋
kanhe 勘合
kefu 客夫
konghui 空回
kouliang yin 口糧銀
kounei 口內
kouwai 口外
kuping 庫平
langshu pi 榔樹皮
langzhong 郎中
lejuan 勒捐
liang 兩
liangtai 糧臺
liangyuan 糧員
liangzhan 糧站
lianxiang 練餉
Liaoxiang 遼餉
lifu 里夫
lijin 釐金
limu 吏目
lingcui 領催
liucun 留存
liuhuang 硫磺
lougui 陋規
luo 騾
luotuo 駱駝
lüqi (*luqi*) 綠旗
lüying bing (*luying bing*) 綠營兵
magan 馬乾
maifu 麥麩
maimaijie 買賣街
maizi 麥子
majia yin 馬價銀
Manbing 滿兵
manfu 蠻夫
Man-Han-tu bing 滿漢土兵
maoqian 毛鉛
maotie 毛鐵
meizhenjian 梅針箭 or 眉針箭
mipiao 米票
mu 畝
mupai 木牌
nanlu 南路
nanxun 南巡
neidachen 內大臣

neige 內閣
neiwufu 內務府
niaoqiang 鳥槍 or 鳥鎗
niaozuichong 鳥嘴銃
niesi 臬司
pao 炮, 礮 or 礮
paotai 炮臺
pentong 噴筒
piao 票
pingyu 平餘
pishan pao 劈山炮
pishan xiaopao 劈山小炮
piyi yin 皮衣銀
pujuan 普蠲
qian 錢
qianfeng 前鋒
qianfeng ying 前鋒營
qianju 錢局
qianzong 千總
qiao 鞘
qima 騎馬
qinchai 欽差
qingke 青稞
qintianjian 欽天監
qintianjian jianfu 欽天監監副
renzi 壬子
shaluoben 沙羅奔
shangshu 尚書
shangshui 商稅
shangyun 商運
shanhou 善後
shaojiu 燒酒
sheng 升
sheng-shu tong-tie 生熟銅鐵
shengtie 生鐵
shengtong 生銅
*shicun*實存
shicun yin 實存銀
shidu 侍讀
shilang 侍郎
shimin 士民
shiquan wugong 十全武功
shou 收
shoubei 守備
shucai zhutan yin 蔬菜燭炭銀
shuli 書吏
shutie 熟鐵
Sichuan li 四川例
sidao 司道
siguan bithesi 司官筆帖式
siku 司庫
siyuan 司員
sui ying 隨營
suiying paoju 隨營炮局
taichangsi 太常寺
taipusi 太僕寺
taixue 太學
taiyiyuan 太醫院
taizhan 臺站
tengpai 藤牌
tiancheng jiangjun pao 天成將軍炮
tianfang qi shui 田房契稅
tiao 條
tidu 提督
tiejiang 鐵匠
tiepao 鐵炮
tingyuan 廳員
tongpan 通判
tongpao 銅炮
tongyi 通譯
tongzhengsi 通政司
tongzhi 同知
tubian 土弁
tubing 土兵
tunfang 屯防
tunlian 屯練
tuntian 屯田
tuoma 馱馬
tusi 土司
waiwei 外委
waixiao 外銷
wang 王
wangfu 王府
wen 文 (coin)
wenguan 文官
wuguan 武官
wula 烏拉 (Mongolian *ulaa*)
wula yun 烏拉運
wuxu 戊戌
xiancheng 縣丞
xiangyong 鄉勇
xiaojixiao 驍騎校 [sic]
xiaozhan 小站
xieke 蟹殼
xieling 協領
xigua pao 西瓜炮
xilu 西路
xin xilu 新西路
xin'an 新案
xingbu 刑部
xingma 行馬
xinhai 辛亥
xinhong zhizhang yin 心紅紙張銀
xingzhuang yin 行裝銀
xinjiang 新疆 'new borderlands'
xinmu bing 新募兵
xinshou 新收
xinyin 薪銀

xuanweishi 宣尉使
xunfu 巡撫
xunjian 巡檢
xushou 續收
xuxiao 續銷
yahang shui 牙行稅
yamen 衙門
yancai 鹽菜
yancai kouliang 鹽菜口糧
yancai yin 鹽菜銀
yanglian 養廉
yanke 鹽課
yanzhengshi 鹽政使
yaodao 腰刀
yaozhan 腰站
yicang 義倉
yigu 義穀
yimin 義民
yingma 營馬
yiwu 夷務
yiwu an 夷務案
yizhan 驛站
youke 遊客
youji 游擊
yubing 餘兵
yuding 餘丁
yuanwailang 員外郎
yuping 餘平
yushitai 御史臺
zakuan 雜款
zaxiang 雜項
zeli 則例
zhandao 棧道
zhang 丈
zhanma 戰馬 (warhorse)
zhanma 站馬 (station horse)
zhanshifu 詹事府
zhanyuan 站員
zhaomo 照磨
zhaopiao 照票
zhenshu 鎮戍
zheng'an 正案
zhengxiang 正項
zhengxiao 正銷
zhengzhan 正站
zhengzhuang yin 整裝銀
zhese 折色
zhifu 知府
zhixian 知縣
zhizhou 知州
zhili ting 直隸廳
zhonglu 中路
zhuangda 壯大
zhuanyun zongju 轉運總局
zhushi 主事
zhun xiao 準銷
zhuxi 鑄息
zongbing 總兵
zongdu 總督
zongli xingying chu 總理行營處
zongrenfu 宗人府
zouxiao 走硝
zuo'er 佐貳
zuoling 佐領
zuoxiang mi 座餉米
zuoxiang yin 座餉銀
zuoza 佐雜

INDEX

www.ingramcontent.com/pod-product-compliance
Lightning Source LLC
LaVergne TN
LVHW020652300926
849161LV00004B/906